HOW TO PASS

FLASH REVISE!

HIGHER
CHEMISTRY

Martin Armitage

HODDER
GIBSON
AN HACHETTE UK COMPANY

Orders: please contact Bookpoint Ltd, 130 Milton Park, Abingdon, Oxon OX14 4SB. Telephone: (44) 01235 827720. Fax: (44) 01235 400454. Lines are open 9.00–5.00, Monday to Saturday, with a 24-hour message answering service. Visit our website at www.hoddereducation.co.uk. Hodder Gibson can be contacted direct on: Tel: 0141 848 1609; Fax: 0141 889 6315; email: hoddergibson@hodder.co.uk

© Martin Armitage 2011

First published in 2011 by
Hodder Gibson, an imprint of Hodder Education,
An Hachette UK Company
2a Christie Street
Paisley PA1 1NB

Impression number   5 4 3 2 1
Year   2013 2012 2011

Cover photo © 1997 Don Farrall, LightWorks Studio/Photodisc/Getty Images; © iStockphoto. com
Illustrations by GreenGate Publishing Services
Typeset in Gill Sans Alternative by GreenGate Publishing Services, Tonbridge, Kent
Printed in Great Britain by CPI Antony Rowe

A catalogue record for this title is available from the British Library

ISBN: 978 1444 120462

Contents

1

Following the course of a reaction

Q1 Name a variable that can be measured to follow the rate of the reaction shown below and state how it could be measured.

$$CaCO_3 + 2HCl \longrightarrow CaCl_2 + CO_2 + H_2O$$

Q2 In a reaction with a gaseous product, $20\,cm^3$ of gas was formed after 10 seconds and $40\,cm^3$ after 50 seconds.

Calculate the average rate, with units, of the reaction during this period of time.

Q3 Insert the missing words:

'The rate of a reaction is _____ to the _____ of the time taken.'

Q4 For the reaction in Q1, explain why there is a steady loss of mass as time passes.

ANSWERS ▶▶

1 Any one of:
- volume of CO_2 produced; a gas syringe could be used to collect CO_2
- mass loss from the reaction; the reaction vessel could be placed on a balance
- pH; the electrode of a pH meter could be placed in the reaction mixture.

2 Rate $= \dfrac{(40 - 20)}{(50 - 10)} = \dfrac{20}{40} = 0.5\,cm^3\,s^{-1}$

3 proportional, reciprocal

4 Carbon dioxide gas is formed in this reaction and is set free. The loss of the gas accounts for the loss of mass.

***Exam* tip:** The rate of a chemical reaction can be found by measuring the change in concentration of a reactant or a product over a period of time. The rate can also be found using a variable that is proportional to a concentration, or from which concentration can be calculated, such as pH. The units of rate must always involve time (usually seconds as s^{-1}).

Factors affecting rate

Q1 Name three factors that can affect the rate of a chemical reaction.

Q2 What is meant by the term 'activation energy'?

Q3 Why does a small increase in temperature produce a large increase in the rate of a chemical reaction?

Q4 Select the correct words to complete the following statement:

'When the temperature at which a chemical reaction is carried out increases, the activation energy (increases / decreases / stays the same) and the number of successful collisions (increases / decreases / stays the same).'

ANSWERS ▶▶

1 temperature, concentration, particle size

2 The activation energy is the minimum kinetic energy needed by colliding particles for reaction to take place.

3 A small increase in temperature causes a large increase in the number of molecules with activation energy.

4 stays the same, increases

Exam tip: In order to react, molecules must first collide. Many collisions do not result in reaction as the molecules are not moving fast enough. They do not have enough kinetic energy. The minimum kinetic energy that molecules must have before collision results in reaction is called the activation energy. An increase in temperature causes an increase in the number of collisions, but, much more importantly, it also increases the number of molecules with the required activation energy.

Activation energy

Q1 Diagrams like the one below are used to show the distribution of kinetic energies in a collection of molecules at two temperatures, T_1 and T_2. The activation energy for the reaction is represented by E_A.

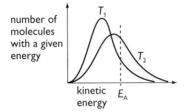

What is represented by the area under each of the curves shown?

Q2 **a** How does T_2 compare with T_1?
  b What happens to the average kinetic energy of the molecules when T_1 is changed to T_2?

Q3 Explain the effect on the rate of reaction when T_1 is changed to T_2.

Q4 What effect would the addition of a catalyst have on the position of the dotted line at E_A?

ANSWERS ▶▶

1 The area under each curve represents the total number of molecules present.

2 **a** T_2 is higher than T_1.
   b The average kinetic energy increases.

3 The rate of reaction increases as more molecules now have the activation energy needed. The number of molecules now having this activation energy is proportional to the area under the curve T_2 to the right of the dotted line. This is much greater than the area under the curve T_1 to the right of the dotted line.

4 A catalyst lowers the activation energy, E_A, and therefore the dotted line would move to the left.

***Exam* tip**: When the temperature is increased, the maximum of the distribution curve moves to the right, but is lower than the maximum of the curve at the lower temperature. This is because the total areas under the curves must remain constant, as the total number of molecules is constant. Possessing at least the activation energy enables molecules to form an activated complex on collision. However, the activated complex can either go on to break down into products or to break down and form the reactants again.

The idea of excess

Q1 In the reaction $Mg + H_2SO_4 \longrightarrow MgSO_4 + H_2$, one mole of magnesium was added to two moles of sulphuric acid. Would all the magnesium react?

Q2 In the reaction $Mg + 2HCl \longrightarrow MgCl_2 + H_2$, five moles of magnesium were added to nine moles of hydrochloric acid. Would all the magnesium react?

Q3 In the reaction $Mg + H_2SO_4 \longrightarrow MgSO_4 + H_2$, 24.3 g of magnesium was added to one litre of $2\,mol\,l^{-1}$ sulphuric acid. Would all the magnesium react?

Q4 A 6 g mass of magnesium was ignited in a sealed container holding 16 g of oxygen gas.

After the reaction stops, what substances will be present in the container?

ANSWERS ▶▶

1 Yes. One mole of magnesium requires only one mole of sulphuric acid to react completely.

2 No. The equation tells us that we need twice as many moles of the acid as moles of magnesium. We need ten moles of acid.

3 One litre of $2 \, mol \, l^{-1}$ acid contains two moles: 24.3 g of magnesium contains one mole, so all the magnesium would react.

4 The equation for the reaction is: $2Mg + O_2 \longrightarrow 2MgO$. Six grams of magnesium is 0.25 mole: 16 g of oxygen is 0.5 mole. That is enough oxygen to react with one mole of magnesium. At the end, there will be magnesium oxide and unreacted oxygen.

***Exam* tip:** To find which reactant is in excess, use the equation to find how many moles of each reactant are involved. Convert given masses into moles (divide the mass by the relative atomic mass (RAM) if it is an element, or by relative formula mass (RFM) if it is a compound). Convert volumes and concentrations into moles. Remember, number of moles is volume in litres multiplied by concentration in mole l^{-1}.

5

Catalysts

Q1 Explain what is meant by the terms:
   a heterogeneous catalyst
   b homogeneous catalyst

Q2 Heterogeneous catalysts adsorb reactant molecules. What does this mean?

Q3 What happens when a heterogeneous catalyst is 'poisoned'?

Q4 Where possible, industrial processes use heterogeneous catalysts rather than homogeneous catalysts. What is the advantage of using a heterogeneous catalyst?

ANSWERS ▶▶

1 **a** A heterogeneous catalyst is in a different physical state from the reactants.

   b A homogeneous catalyst is in the same physical state as the reactants.

2 Adsorption means that bonds form between the reactant molecules and the atoms on the surface of the catalyst.

3 A 'catalyst poison' bonds more tightly to the catalyst surface than the reactant molecules do. As a result, the reactant molecules cannot be adsorbed.

4 Industrial processes tend to favour heterogeneous catalysts, many of which are solids involving transition metals, as it is easier to separate the products from a solid catalyst.

***Exam* tip:** Biological catalysts are called enzymes, and these are also used in industry as well as occurring in living organisms. Enzymes are denatured at higher temperatures (their structures are altered) and therefore must be used at lower temperatures.

Effect of concentration on reaction rate

In the Prescribed Practical Activity (PPA) 'Effect of Concentration Changes on Reaction Rates', to find the effect of varying the concentration of iodide ions on the rate of reaction between hydrogen peroxide and an acidified solution of potassium iodide, hydrogen peroxide oxidises iodide ions to give iodine. The iodine is then converted back to iodide by thiosulphate ions until all the thiosulphate has been used up, when free iodine begins to form.

The following mixtures are used.

mixture	H_2SO_4 (cm³)	thiosulphate (cm³)	starch (cm³)	potassium iodide (cm³)	water (cm³)
1	10	10	1	25	
2	10		1	20	
3	10		1	15	
4	10		1	10	
5	10		1	5	

A 5 cm³ volume of hydrogen peroxide is then added by syringe to each mixture. The time taken for free iodine to form is noted.

Q1 How can the moment at which free iodine forms be detected?

Q2 What are the missing figures in the thiosulphate column?

Q3 What are the missing figures in the water column?

Q4 How is the rate of the reaction for each mixture found?

ANSWERS ▶▶

1 When free iodine begins to form, it reacts with the starch and the mixture turns blue–black.

2 The missing figures are all $10\,cm^3$. This means that, in each experiment, the same amount of iodine has to form before the mixture turns blue–black. The fact that the same amount of iodine has to form makes the comparison between mixtures a fair one.

3 The missing figures descending the water column are 0, 5, 10, 15 and 20.

Using these volumes of water ensures that the total volume of liquid in each experiment is the same – that is, $51\,cm^3$. If these volumes of water were not added, each experiment would have a different volume and the concentration of the hydrogen peroxide and acid would be different in each experiment.

4 The rate is calculated as $\frac{1}{t}$, where t is the time taken for the mixture to become blue–black

***Exam* tip:** It is valid to calculate rate as $\frac{1}{t}$ as a result of the design of the experiment. Since the same volume of thiosulphate solution is used in each experiment, the same amount of free iodine has to form before the starch changes colour. This means that each mixture is allowed to react to the same extent before timing is stopped.

Potential energy diagrams (1)

Q1 Draw the shape of a potential energy diagram for a reaction for which the activation energy is $100\,kJ\,mol^{-1}$ and the enthalpy change is $-200\,kJ\,mol^{-1}$.

Q2 Is this reaction exothermic or endothermic?

Q3 Calculate the activation energy of the reverse reaction.

Q4 Explain why an endothermic reaction (a reaction that takes in energy) results in a fall in temperature.

ANSWERS ▶▶

1

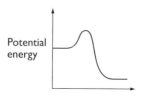

Potential energy

2 Exothermic, as the products have less potential energy than the reactants.

3 Activation energy is 300 kJ mol^{-1}. The activation energy for the reverse reaction is ΔH for the reverse reaction added to the activation energy for the forward reaction.

4 In an endothermic reaction, heat energy is taken in from the surroundings by the reactants as they turn into products. This heat energy becomes part of the chemical energy of the products and does not make them hotter.

***Exam* tip:** Enthalpy changes in reactions are measured in kJ mol^{-1}. The symbol for an enthalpy change is ΔH, which has a negative value for an exothermic reaction and a positive value for an endothermic reaction. Reactants and products usually contain different amounts of chemical energy. When reactants turn into products the difference in chemical energy is made up by either giving out or taking in heat energy. The enthalpy change for a reverse reaction is numerically identical to that for the forward reaction, but the sign is reversed.

Potential energy diagrams (2)

Q1 What is meant by the term 'activated complex'?

Q2 Describe how a heterogeneous catalyst speeds up a reaction.

Q3 Draw a potential energy diagram for an exothermic reaction, showing both catalysed and uncatalysed processes.

Q4 What effect does the enthalpy change for the reaction have on the rate of reaction?

ANSWERS))

1 An activated complex is an unstable arrangement of atoms formed when the molecules with the highest kinetic energy collide. The activated complex *can* break down to form the products of the reaction.

2 Reactant molecules are adsorbed on the surface of the heterogeneous catalyst where they form an activated complex with a lower energy than when a catalyst is not used.

3

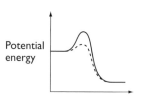

Potential energy

The dotted line represents the catalysed process.

4 The enthalpy of reaction has no effect on the rate of the reaction. The rate depends entirely on the activation energy. The higher the activation energy, the slower the reaction.

***Exam* tip:** The activated complex is an unstable arrangement of atoms at the maximum of the potential energy barrier. Catalysts allow an alternative complex to form; this alternative complex requires less energy to form. An activated complex can break down to form products, but it can also break down to the reactant molecules again.

Enthalpy changes (1)

Q1 What is meant by the term 'enthalpy of combustion'?

Q2 ΔH for the reaction $2C + O_2 \longrightarrow 2CO$ does not represent the enthalpy of combustion of carbon. Why not?

Q3 When 0.1 mole of ethane was burned completely, 156 kJ of energy were released.

What is the enthalpy of combustion of ethane?

Q4 The enthalpies of combustion of ethanol (C_2H_5OH) and dimethyl ether (CH_3OCH_3) are different. This is because:

A their boiling points are different
B their RFMs are different
C their intramolecular bonding is different
D their combustion products are different.

ANSWERS ▶▶

1   It is the enthalpy change when one mole of a substance is burned completely in oxygen.

2   Since the carbon is burning to form carbon monoxide, rather than carbon dioxide, the combustion is incomplete.

3   $\Delta H = -1560\,\text{kJ}\,\text{mol}^{-1}$. The units and the negative sign are essential.

4   **C**. These compounds are isomers of each other so they have the same RFM. Both burn to form carbon dioxide and water. The molecules have different structures, which means that the pattern of bonding within the molecule (intramolecular) is different. This results in the different values for enthalpy of combustion.

Exam tip: Enthalpies of combustion are often found by using the heat produced to warm a known mass of water. The amount of heat released is $cm\Delta T$ where $c = 4.18\,\text{kJ}\,\text{kg}^{-1}\,^{\circ}\text{C}^{-1}$ and m is the mass of water heated in kg. ΔT is the rise in temperature of the water caused by the heating. Enthalpies of combustion for most common substances are negative as the reactions are exothermic.

Enthalpy changes (2)

Q1 What is meant by the term 'enthalpy of solution'?

Q2 Exactly 0.1 mole of an ionic compound was dissolved in one litre of water. As it dissolved, the temperature fell by 2 °C. How much heat was taken in by the compound?

Q3 What is ΔH for the process in Q2?

Q4 The process in Q2 is endothermic, as the temperature fell. This means that bonds are being broken. Suggest which bonds are being broken.

ANSWERS ▶▶

1 The enthalpy of solution of a substance is the enthalpy change when one mole of the substance dissolves completely in water.

2 Heat taken in $= cm\Delta T = 4.18 \times 1 \times 2 = 8.36\,kJ$.

3 $\Delta H = +83.6\,kJ\,mol^{-1}$. The positive sign is essential, as this is an endothermic reaction. The compound took in heat from the water as it dissolved. It is necessary to multiply 8.36 by 10, as only 0.1 mole of compound was dissolved in water, but the enthalpy of solution refers to one mole of compound.

4 The compound is ionic. It consists of a lattice of ions held together by electrostatic forces. When it dissolves, the ions are separated from each other. This means that the ionic bonds are being broken.

***Exam* tip:** Enthalpies of solution are always found experimentally using the $cm\Delta T$ expression. Many enthalpies of solution are endothermic (ΔH positive) and some are exothermic (ΔH negative). When an ionic compound dissolves in water, two processes take place. Water molecules form bonds with the ions (exothermic) and the lattice breaks down (endothermic).

Enthalpy changes (3)

Q1 What is meant by the term 'enthalpy of neutralisation'?

Q2 Why is the enthalpy of neutralisation always exothermic (ΔH is negative)?

Q3 The enthalpy of neutralisation of sulphuric acid is $-57\,kJ\,mol^{-1}$. When one mole of sulphuric acid is neutralised, $114\,kJ$ are released. Why?

Q4 Suggest why the enthalpy of neutralisation of a weak acid (such as ethanoic acid, CH_3COOH) is $1.2\,kJ\,mol^{-1}$ less than that of a strong acid (such as HCl).

ANSWERS ▶▶

1 The enthalpy of neutralisation of an acid is the enthalpy change when the acid is neutralised to form one mole of water.

2 It is exothermic, as it always involves H^+ ions from the acid bonding with OH^- ions from the alkali. Since bonds are formed, the process is exothermic so ΔH is negative.

3 The enthalpy of neutralisation refers to the formation of one mole of water. Sulphuric acid, H_2SO_4, forms two moles of hydrogen ions and therefore two moles of water form when it is neutralised.

4 Since ethanoic acid is a weak acid, it is not completely dissociated into ions. To break the bond holding the hydrogen ion to the rest of the molecule, energy is required. This means that less energy is released overall in the neutralisation.

Exam tip: When one mole of H^+ ions combines with one mole of OH^- ions, approximately 57 kJ are released. This figure is approximately correct for all strong acids (acids that break down completely into ions). The figure is lower for weak acids.

Enthalpy changes (4)

Q1 In the PPA 'Enthalpy of Combustion' in which ethanol is burned at the wick of a small burner, in order to heat water in a copper container what are the experimental measurements that must be made?

Q2 Why is a copper container preferable to a glass container?

Q3
a What are the main sources of error in this experiment?
b What can be done to reduce the sources of error?

Q4 Why should a cap be kept over the wick of the burner except when the ethanol is actually being burned?

ANSWERS ▶▶

1 • mass of the burner, containing its fuel, before and after combustion
  • temperature of the water before and after combustion
  • mass of water.

2 Copper is a much better conductor of heat than glass.

3 a The main sources of error are heat loss and incomplete combustion of the ethanol.

  b A draught shield can be used to ensure that as much as possible of the heat from the flame reaches the copper container.

4 When the cap is removed, ethanol can be lost by evaporation.

***Exam* tip:** When asked what measurements are required in this PPA, it is necessary to give the answers in Answer 1. These are experimental data. It is not correct to say 'change in mass of burner' and 'rise in temperature of the water' as these are calculations, not observations. Evidence for incomplete combustion of ethanol is seen from the fact that the bottom of the copper container becomes covered with soot, indicating that some of the ethanol has burned to form carbon, not carbon dioxide.

The periodic table (1)

Q1 What is the basis of **a** the left to right and **b** the vertical arrangements of the elements in the periodic table?

Q2 What name is given to **a** a horizontal row of elements and **b** a vertical column of elements in the periodic table?

Q3 Explain the general trend in melting point and boiling point as Group 7, the halogens, is descended.

Q4 Suggest why the melting points and boiling points of sodium, magnesium and aluminium (elements 11, 12 and 13) increase from sodium to aluminium.

ANSWERS

1 **a** increasing atomic number
  b similar chemical properties

2 **a** period
  b group

3 The melting points and boiling points increase as we go down group 7.

  This is because the weak intermolecular forces (Van der Waals) increase as the molecule becomes larger, so more energy is needed to separate molecules.

4 Sodium has one outer electron, magnesium has two and aluminium three. This contributes to stronger metallic bonding from sodium to aluminium.

***Exam* tip:** The modern periodic table is based on the work of the Russian chemist Mendeleev, who arranged the elements in order of increasing atomic weights, with elements showing similar properties placed in the same vertical column. Where appropriate, he ignored the atomic weights in favour of placing similar elements in the same column, and left gaps where no element, known at that time, would fit.

The periodic table (2)

Q1 Describe and explain the trend in atomic radius of elements as a group of the periodic table is descended.

Q2 Describe and explain the trend in atomic radius as we go from left to right across a period of the periodic table.

Q3 Insert the missing words. 'Aluminium (atomic number 13) is denser than magnesium (atomic number 12). This is because the radius of aluminium atoms is _____ than the radius of magnesium atoms, while the mass of aluminium atoms is _____ than the mass of magnesium atoms.'

Q4 The element tin has two different forms. Grey tin is a non-metal with a structure like that of silicon. White tin is a metal. By considering the position of tin in the periodic table, suggest why it is not surprising that tin has these two forms.

ANSWERS ▶▶

1 The atomic radius increases down a group owing to each atom having one additional electron shell. The nuclear attraction for this outer shell is lower as a result of screening by the inner electron shells.

2 The atomic radius decreases from left to right across a period, because the nuclear charge increases, resulting in a greater force of attraction between the nucleus and the electrons, drawing the electron shells closer to the nucleus.

3 less, greater

4 Tin, in group 4, lies on the boundary line between metals and non-metals in the periodic table. Elements on this boundary often show properties of both metals and non-metals – for example, graphite conducts electricity and silicon is a semiconductor. In the case of tin, two separate forms exist.

***Exam* tip:** Atomic radius decreases along a period mainly because of increased nuclear attraction for the outer electrons, and increases down a group, mainly because of the additional electron shell. As each group is descended, there is a large increase in the mass of the atoms. As a result, densities, melting points and boiling points tend to increase as a group is descended, although this is not always the case.

Ionisation energy (1)

Q1 What is meant by the term 'first ionisation energy'?

Q2 Explain the trend in ionisation energies across a period.

Q3 Explain the trend in ionisation energies down a group.

Q4 Why is the second ionisation energy always larger than the first ionisation energy?

ANSWERS ▶▶

1 It is the energy required to remove one mole of electrons from one mole of gaseous atoms of an element. The first ionisation energy refers to the outermost electron, or one of the outermost electrons.

It is the energy change for the process $E(g) \longrightarrow E^+(g) + e^-$.

2 Ionisation energies increase across a period, because the atoms have more protons, and since the atoms become smaller, the outer electrons are closer to the nucleus.

3 Ionisation energies decrease down a group, because the outermost electrons are further from the nucleus, and are screened from the nucleus by inner electron shells.

4 Once an electron has been removed from an atom, the atom now has a positive charge. It is harder to remove an electron from a positive ion than it is to remove it from a neutral atom.

Exam tip: The second and third ionisation energies are the amounts of energy required to remove a second and third mole of electrons. The second ionisation energy is larger than the first because the atom now has a 1+ charge, and the third is larger than the second as the atom now has a 2+ charge. Note that the definition of ionisation energy is found on page 10 of the SQA Data Booklet.

Ionisation energy (2)

Q1 Which of the following equations represents the first ionisation energy of fluorine?

A $F^-(g) \longrightarrow F(g) + e^-$
B $F^-(g) \longrightarrow \frac{1}{2}F_2(g) + e^-$
C $F(g) \longrightarrow F^+(g) + e^-$
D $\frac{1}{2}F_2(g) \longrightarrow F^+(g) + e^-$

Q2 How is the energy change for the following process calculated?
$Mg(g) \longrightarrow Mg^{2+}(g) + 2e^-$

Q3 Why does helium not have a third ionisation energy?

Q4 The first three ionisation energies of beryllium are 905, 1770 and 14800 kJ mol^{-1}, respectively. Why is there such a large increase from the second to the third value?

ANSWERS ▶▶

1   **C**. The first ionisation energy of fluorine refers to a neutral atom, not a fluoride ion, as in **A** and **B**, nor a fluorine molecule, as in **D**.

2   It is the sum of the first and second ionisation energies.

3   Helium, with Atomic Number 2, has only two electrons. It cannot lose a third electron.

4   The third electron is removed from an inner electron shell, which is full and closer to the nucleus.

Exam tip: As more electrons are removed from the atom, the energy required increases to a set pattern. The energy required to form an $n+$ ion is the sum of the first n ionisation energies. When removal of an electron from a new shell takes place, there is a very large increase in the energy required as the electron is in a stable shell and is closer to the nucleus.

Note that the equation describing the first ionisation energy refers to separate isolated atoms of the element. The state symbol (g) must always be included when writing the equation.

Electronegativity and bonding

Q1 What is meant by the term 'electronegativity'?

Q2 What are the general trends in electronegativity
   a across a period
   b down a group?

Q3 Where, in the periodic table, are the elements with the
   a highest
   b lowest

   electronegativities?

Q4 What type of bonding results when two atoms have electronegativities that are:
   a low and identical
   b high and identical
   c very different
   d high but not identical?

ANSWERS ▶▶

1 Electronegativity is a measure of the attraction an atom has for the electrons in a bond in which the atom is involved.

2 **a** increasing
b decreasing

3 **a** The elements with the highest electronegativities are at the top right of the table: nitrogen, oxygen and fluorine.
b The elements with the lowest electronegativities are at the bottom left of the table: rubidium, caesium and francium.

4 **a** metallic
b pure covalent
c ionic
d polar covalent

Exam tip: In general, metals have low electronegativities and non-metals have high electronegativities. This is consistent with the idea that the bond between a metal and a non-metal is ionic and the bond between two non-metals is covalent. Pure covalent bonding exists almost exclusively in non-metal elements, while all other covalent bonds are almost certain to be polar to some extent. The further apart two elements are in the periodic table, the more ionic character their bond will have.

Bonding

Q1 In metallic bonding, what term is used to describe electrons that are free to move throughout the metallic lattice?

Q2 Insert the missing words: 'Atoms in a covalent bond are held together by _____ forces of attraction between _____ charged nuclei and negatively charged _____ electrons.'

Q3 a What forces hold ions together in an ionic bond?
  b How would the ionic bond strength in NaCl be expected to compare with that in MgO?

Q4 The electronegativities of Li, Cs, F and I are, respectively, 1.0, 0.8, 4.0 and 2.6.

Which compound shows the greatest covalent character: LiF, LiI, CsF or CsI?

ANSWERS ▶▶

1 delocalised

2 electrostatic, positively, shared

3 a In an ionic bond, the ions are held together by electrostatic forces.
   b In NaCl, the ions both have a single charge. In MgO, each ion is doubly charged. The attraction between doubly charged ions is much stronger than that between singly charged ions.

4 Lithium iodide. This involves the smallest difference in electronegativity between the group 1 element and the group 7 element.

***Exam* tip:** The strength of ionic bonding depends largely on the charges of the ions involved – the higher the charge, the stronger the force of attraction. Compounds between groups 1 and 7 have much weaker ionic bonding than those between groups 2 and 6.

With the exception of metallic bonding, most bonds have both ionic and covalent character, depending on the difference in the electronegativities of the atoms involved. If the electronegativities are identical, then the bond is pure covalent. A polar bond can be thought of as a covalent bond with some ionic character. The greater the difference between two atoms in electronegativity the greater the degree of bond polarity, resulting in greater ionic character and less covalent character.

Intermolecular forces of attraction (1)

Q1 What causes Van der Waals' forces?

Q2 **a** How is the strength of Van der Waals' forces related to the size of molecules?
b How is the strength of Van der Waals' forces related to the ability of the molecules to get close together?

Q3 How does the strength of Van der Waals' forces compare with that of other kinds of bonding?

Q4 Use the idea of Van der Waals' forces to explain why branched hydrocarbons have lower boiling points than their straight-chain isomers.

ANSWERS »

1 Van der Waals' forces between molecules are caused by the attraction of a temporary dipole in one molecule to an induced dipole in a neighbouring molecule.

2 a The larger the molecules, the stronger the Van der Waals' forces between them.

   b The closer the molecules can approach, the stronger the Van der Waals' forces.

3 Van der Waals' forces are by far the weakest kind of bonds.

4 The molecules of branched hydrocarbons cannot approach each other as closely as those of straight-chain hydrocarbons. The Van der Waals' forces are therefore weaker and the hydrocarbons have lower boiling points.

Exam tip: It is thought that a neutral molecule can become a dipole – that is, have a positive end and a negative end – as a result of electrons being momentarily unevenly distributed in the molecule. This momentary polarity affects neighbouring molecules. The positive charge attracts the electrons of a neighbouring molecule, making one end of it slightly negatively charged and, as a result, the opposite end of the neighbouring molecule becomes slightly positively charged. The two molecules are therefore attracted to each other as a result of this temporary electrostatic effect.

Intermolecular forces of attraction (2)

Q1 What causes a bond to be a permanent dipole?

Q2 Molecules containing bonds that are permanently polar may not themselves be polar. Why not?

Q3 All of the following molecules contain polar bonds: H_2O, CCl_4, CO_2, CH_4. Only one of these molecules has overall polarity. Explain which one of these molecules is polar.

Q4 Boron trifluoride, BF_3, is a flat molecule, with a boron atom in the centre, linked to three fluorine atoms. It is non-polar.

Ammonia, NH_3, is a polar molecule. What feature of its structure makes it polar?

ANSWERS ▶▶

1 A bond is a permanent dipole when the two atoms involved in the bond have different electronegativity values. The end of the bond with the more electronegative atom is negative with respect to the other end of the bond.

2 In order for a molecule to be polar, it must contain polar bonds, but the molecule must not be too symmetrical otherwise the effects of the polar bonds cancel out.

3 Only H_2O is polar. The other molecules are too symmetrical, resulting in bond polarity being cancelled out. The water molecule is V shaped, with the oxygen atom at the apex of the V and the hydrogens at the ends. This means that the hydrogens carry a slight positive charge and the oxygen a slight negative charge.

4 Ammonia has a pyramidal structure. The nitrogen atom is at the top of the pyramid and the hydrogens are at the foot. As a result, the molecule has a negative end (the nitrogen) and a positive end (the hydrogens).

***Exam* tip:** Highly symmetrical molecules like CO_2, which is linear (O=C=O), and tetrahedral molecules where all four atoms attached to carbon or silicon are identical cannot be polar even though the interatomic bonds are polar. The polar bonds in such molecules are described as being 'symmetrically opposed' to each other.

Bonding and structure in elements

Q1 Describe the structure of a metal and explain how it is held together as a solid.

Q2 What are the two possible structures for a non-metallic element that is not a noble gas?

Q3 Name an element that can have either of the structures referred to in Q2.

Q4 Explain why the bonding between atoms in a non-metallic element must be pure covalent.

ANSWERS ▶▶

1   A metal comprises a lattice of positive ions. Between the positive ions there is a 'sea' of delocalised electrons. The attraction of the electrons for the positive ions holds the positive ions together and the attraction of the ions for the electrons prevents the loss of electrons by the solid metal.

2   Non-metallic elements can have either a discrete molecular structure or a giant lattice structure.

3   carbon

4   In a non-metallic element the bonding must be pure covalent. Since it is an element, all the atoms must be identical. If they are identical, they have identical electronegativities. As a result, the atoms have identical attractions for electrons, which are therefore shared equally between them.

***Exam* tip:** Elements can have only a metallic structure (if they are metals) or covalently bonded structures (if they are non-metals). Covalent networks involve non-metal elements with a valency of four although non-metal elements with a valency of three can also form covalent networks. Non-metals with a valency of one or two form covalent molecules between which there are weak Van der Waals' forces. Noble gases are monatomic; they exist as single, isolated atoms.

Structure in elements and compounds

Q1 Name the three different structures in which the element carbon can exist.

Q2 What types of structures are found in each of the following compounds:

a lithium iodide
b iodine monochloride
c silicon dioxide?

Q3 Substance X is melted and an electric current is passed through it. This results in the breakdown of the substance into two elements. The type of structure present in substance X is:

A covalent molecular
B covalent network
C ionic
D metallic

Q4 Name the element that exists as molecules containing eight atoms.

ANSWERS ▶▶

1   Carbon can exist as diamond, graphite and fullerenes. Diamond and graphite are covalent networks, while fullerenes are discrete covalent molecules.

2   **a**   ionic
    b   covalent molecular
    c   covalent network

3   **C** (ionic). Covalent substances cannot conduct electricity. This rules out **A** and **B**. When a current is passed through a metal **D**, it is not broken down. This leaves option **C**, the ionic substance. When a current is passed through a molten ionic substance, electrolysis takes place.

4   sulphur

***Exam* tip:** You are expected to know the structures of the first 20 elements. If the element is a metal it has a metallic structure. The metals are Li, Be, Na, Mg, Al, K and Ca. Noble gases are monatomic; He, Ne and Ar exist as separate atoms. B, C and Si exist as covalent networks. The rest exist as covalent molecules. C can also form fullerene molecules.

Compounds can be ionic if the electronegativities of the combining elements are sufficiently different. This is often the case when a metal combines with a non-metal. Compounds can also exist as covalent molecules or covalent networks.

Properties of elements and compounds (1)

Q1 Why is almost every ionic compound solid?

Q2 Butane (C_4H_{10}) and propanone (C_3H_6O) both have a relative formula mass of 58. Explain why butane is a gas at room temperature, while propanone is a liquid.

Q3 Explain why carbon in the form of diamond is very hard, while carbon in the form of graphite is so soft.

Q4 Explain why carbon dioxide is a gas, while silicon dioxide is a solid.

ANSWERS ▶▶

1 There are strong electrostatic forces between ions of opposite charge. These forces hold the ions tightly together in a stable lattice, which requires a lot of energy to be broken down.

2 In propanone, the $C{=}O$ bond is polar because carbon and oxygen have quite different electronegativities. Because the molecule is not too symmetrical, it is polar so that propanone contains some intermolecular bonding resulting in a higher boiling point than butane.

3 In diamond, each carbon atom is bonded to four others in a three-dimensional network. In graphite, each carbon atom is bonded to three others, forming sheets that can easily slide over each other.

4 Carbon dioxide consists of discrete CO_2 molecules, while SiO_2 forms a three-dimensional covalent network.

***Exam* tip**: The physical properties of substances depend on the type of bonding and structure within the substance. In turn this depends largely on electronegativity values. If there are big differences in electronegativities the bonding will be ionic or involve polar molecules, often resulting in high melting and boiling points. Otherwise non-polar molecules tend to form, resulting in gases, liquids or low melting solids. Covalent networks have high melting and boiling points.

Properties of elements and compounds (2)

Q1 There are several types of bond: covalent, hydrogen bonding, ionic and Van der Waals.

Arrange these in increasing order of strength.

Q2 Explain why PCl_3 is a liquid at room temperature, but NaCl is a solid.

Q3 Explain why PCl_3 is soluble in $CCl_4(l)$ but NaCl is not.

Q4 Phosphorus forms another chloride, PCl_5, phosphorus pentachloride. It can be represented as $[PCl_4^+][PCl_6^-]$. Explain whether it is likely to exist as a solid or a liquid.

ANSWERS ▶▶

1 Van der Waals' bonds are weaker than hydrogen bonding. Covalent and ionic bond may be thought of as equally strong, and stronger than either of Van der Waals or hydrogen bonding.

2 The electronegativity difference between phosphorus and chlorine is small enough for covalent bonding to occur. The electronegativity difference between sodium and chlorine is large, so ionic bonds form. This means that phosphorus trichloride exists as molecules, held to each other by weak Van der Waals' bonds. Sodium chloride forms as an ionic lattice.

3 Like PCl_3, CCl_4 is covalent. It is non-polar as it has a highly symmetrical structure (tetrahedral). In general, covalent compounds tend to dissolve in non-polar solvents, and ionic compounds tend to dissolve in polar solvents, like water.

4 As this compound can exist in an ionic form, it is a solid at room temperature.

Exam tip: Most ionic compounds are soluble to some extent in water. Compounds where both ions have a single charge tend to be very soluble. If both ions have a double charge (like MgO) the electrostatic attraction between ions is much stronger and they are usually much less soluble. The solubility of ionic compounds in polar solvents and of covalent compounds in non-polar solvents is sometimes expressed as 'like dissolves like'.

Properties of compounds

Q1 Name the three elements (apart from hydrogen) that are involved in hydrogen bonding.

Q2 Explain why ice is less dense than water.

Q3 HCl (RFM 36.5), HBr (RFM 81) and HI (RFM 128) are all gases at room temperature. Explain why HF (RFM 20) is a liquid although its molecules are much smaller than those of the other group 7 hydrides.

Q4 Explain how the viscosity of 1,2-dihydroxyethane (CH_2OHCH_2OH) will compare with that of methanol, CH_3OH.

ANSWERS ▶▶

1 nitrogen, oxygen, fluorine

2 Hydrogen bonding between H_2O molecules in water holds the molecules apart in an open structure. The molecules are further apart than they are in liquid water.

3 Hydrogen bonding is present in HF, resulting in a polymer-like structure.

4 1, 2-dihydroxyethane has two hydroxyl groups that can take part in hydrogen bonding. Methanol has only one. The greater opportunity for hydrogen bonding between molecules makes 1, 2-dihydroxyethane more viscous than methanol.

Exam tip: Hydrogen bonding has a major influence on the properties of compounds where it occurs. It affects boiling points, melting points, viscosity and the ability of the compound to act as a solvent, especially for solutes that are ionic or polar.

Hydrogen bonding can be intermolecular as in all the above examples. Intramolecular hydrogen bonding is partly responsible for the structure of protein molecules.

Avogadro's constant and gas volumes (1)

Q1 How many atoms are contained in two moles of carbon?

Q2 How many molecules of methane are contained in 12 litres of the gas?

Q3 In the following reaction:

$$N_2(g) + 3H_2(g) \longrightarrow 2NH_3(g)$$

what volume of NH_3 will form from the complete reaction of $20\,cm^3$ of $N_2(g)$?

Q4 Which of the following gas samples has the same volume as $7\,g$ of CO (RFM 28)?

A $8\,g$ of CH_4 (RFM 16) **B** $1\,g$ of He (RAM 4)
C $1.0\,g$ of H_2 (RFM 2) **D** $35.5\,g$ of Cl_2 (RFM 71)

ANSWERS ▶▶

1 One mole contains 6×10^{23} atoms. Two moles therefore contain 1.2×10^{24} atoms.

2 One mole of gas at room temperature and pressure (RTP) occupy about 24 litres, so 12 litres contain 0.5 mole. This is $0.5 \times 6 \times 10^{23} = 3 \times 10^{23}$ molecules.

3 The reacting volumes of gases are directly proportional to the numbers of moles taking part. Therefore, $40\,cm^3$ of NH_3 will form from $20\,cm^3$ of N_2.

4 **B**. Samples of gas containing the same numbers of moles have equal volumes.

Here we have $\dfrac{7}{28} = 0.25$ moles of CO.

In **A** there is 0.5 mole; in **B** there is 0.25 mole; in **C** there is 0.5 mole; in **D** there is 0.5 mole.

Exam tip: Gas volume calculations are based on the fact that equal volumes of gases contain equal numbers of moles of gas. This is because it is found that one mole of any gas at RTP occupies the same volume, around 24 litres. Therefore volumes and numbers of moles must be directly proportional to each other. It follows also that equal volumes of gases at RTP must contain equal numbers of molecules.

Avogadro's constant and gas volumes (2)

Q1 How many atoms are contained in 0.5 mole of propane?

Q2 How many molecules of propane are contained in 48 litres of the gas?

Q3 In the following reaction:

$$C_3H_8(g) + 5O_2 \longrightarrow 3CO_2 + 4H_2O$$

what volume of gas will form from the complete combustion of 20 cm^3 of $C_3H_8(g)$,

a when the products are above 100 °C

b when the products are at room temperature?

Q4 Which of the following gas samples has the same volume as 22 g of C_3H_8 (RFM 44)?

A 142 g of Cl_2 (RFM 71) **B** 1 g of He (RAM 4)
C 2.0 g of H_2 (RFM 2) **D** 8 g of CH_4 (RFM 16)

ANSWERS ▶▶

27 Avogadro's constant and gas volumes (2)

1 One mole contains 6×10^{23} atoms. Therefore 0.5 mole contains 3×10^{23} atoms.

2 One mole of gas at RTP occupy about 24 litres, so 48 litres contains two moles. This represents $2 \times 6 \times 10^{23}$ molecules $= 1.2 \times 10^{24}$ molecules.

3 The reacting volumes of gases are directly proportional to the numbers of moles taking part. Therefore, $20 \, cm^3$ of C_3H_8 will form $60 \, cm^3$ of CO_2 and $80 \, cm^3$ of water vapour if the temperature is above $100 \, ^\circ C$.

  a $140 \, cm^3$ ($CO_2 + H_2O$)

  b $60 \, cm^3$ (CO_2 only)

4 **D**. Gases containing the same numbers of moles have equal volumes. There are $\dfrac{22}{44} = 0.5$ mole of C_3H_8.

In **A** there are 2 moles; in **B** there is 0.25 mole; in **C** there is 1 mole; in **D** there is 0.5 mole.

***Exam* tip:** In calculations involving reacting gases (especially combustion of hydrocarbons) where water is a product, it is necessary to be clear about the temperature of the products. Above $100 \, ^\circ C$ water exists as a gas and will occupy about 24 litres per mole. Below $100 \, ^\circ C$, water will be a liquid and have negligible volume.

Petrol (1)

Q1 Petrol can be made by the 'reforming' of naphtha. What is meant by the term 'reforming'?

Q2 Apart from straight-chain alkanes, what three other types of molecule are produced by the reforming process?

Q3 What is the usual term used for 'auto-ignition' in a car engine?

Q4 At one time lead compounds were added to petrol to prevent auto-ignition.

Now the composition of petrol is modified to reduce auto-ignition. In what way is the composition modified?

ANSWERS ▶▶

1   Reforming is a process carried out on hydrocarbon molecules that alters the structure of the molecule without changing the number of carbon atoms in the molecule.

2   Reforming produces branched, cyclic and aromatic hydrocarbons.

3   knocking

4   Lead-free petrol contains increased quantities of branched, cyclic and aromatic hydrocarbons and smaller quantities of straight-chain hydrocarbons.

Exam tip: Petrol is a mixture of hydrocarbons: straight chain, branched, cyclic and aromatic. Reforming is used to generate the last three, and involves the use of heat and suitable catalysts. Different blends of petrol are used at different times of year. For example, in winter the amount of smaller hydrocarbon molecules, which are more easily vapourised, is boosted in the mixture.

Petrol (2)

Q1 Which of the following compounds is most likely to result from the reforming of heptane?

A octane   **B** hex-2-ene

C methylbenzene   **D** cyclobutane

Q2 Which potential pollutant is not the result of incomplete combustion?

A carbon monoxide   **B** carbon

C octane   **D** nitrogen dioxide

Q3 What type of metal is used in a car's catalytic converter?

Q4 'Oxygenates' are oxygen-containing compounds added to petrol to improve its performance. Name a commonly added oxygenate.

ANSWERS ⟩⟩

1 **C**. Since reforming alters the structure of the molecule without changing the number of carbon atoms in the molecule, methylbenzene (toluene) is the most likely product of reforming. Octane has eight carbon atoms, hex-2-ene has six and cyclobutane has four.

2 **D**. Nitrogen dioxide is not the result of incomplete combustion. Octane would be an unburned hydrocarbon, while carbon and carbon monoxide are the result of partial combustion.

3 The catalyst involves precious transition metals such as platinum and rhodium.

4 Ethanol is frequently added to petrol.

***Exam* tip:** A 'three-way' catalyst system oxidises carbon monoxide and unburned hydrocarbons, and reduces nitrogen oxides to nitrogen and oxygen. Such systems require careful control of the air/petrol mix that goes to the cylinders and this is achieved using oxygen sensors. Nitrogen oxides are not produced in large amounts by a 'lean-burn' engine, so this kind of catalyst system is not necessary.

Platinum/rhodium catalysts work at temperatures lower than those required for a pure platinum catalyst.

Oxygenates like ethers, R—O—R' (where R is a hydrocarbon chain) are also added to petrol to improve performance. The most commonly used is methyl tertiary-butyl ether (MTBE), formula $(CH_3)_3COCH_3$.

Alternative fuels

Q1 Petrol is not a renewable fuel. Ethanol, which is a renewable fuel, can be added to petrol. Why can ethanol be called 'renewable'?

Q2 Give one advantage and one disadvantage of using methanol as a fuel in car engines.

Q3 What is the main advantage of using hydrogen as a fuel?

Q4 Biogas is formed under anaerobic conditions by the fermentation of biological materials.

  a What does 'anaerobic' mean?
  b What is the main constituent of biogas?

ANSWERS))

1 Ethanol can be formed by the fermentation of cane sugar. Since new crops of cane sugar can be grown each year, it is renewable, and so is the ethanol made from it.

2 An advantage of using methanol is that it is less flammable than petrol and therefore safer. However, methanol can absorb water and this makes it a potential cause of corrosion in the car engine.

3 The only product of the combustion of hydrogen is water. Hydrogen is virtually pollution-free as a fuel.

4 **a** 'Anaerobic' means 'in the absence of air'.
   b methane

***Exam* tip:** The main advantage of hydrogen as a fuel for cars is the lack of pollution. The hydrogen could be combined with oxygen in a fuel cell, which uses the reaction between hydrogen and oxygen to generate electricity, which in turn powers the car. Hydrogen would have to be made in large amounts, possibly by electrolysis, and hydrogen filling stations would be required.

Hydrocarbons

Q1 Name the following hydrocarbon:

$$CH_3 - CH_2 - CH_2 - CH - CH_3$$
$$\quad\quad\quad\quad\quad\quad\quad | $$
$$\quad\quad\quad\quad\quad\quad CH_3$$

Q2 Name the following hydrocarbon:

$$CH_3 - CH_2 - CH = CH_2$$

Q3 Name the following hydrocarbon:

$$\quad\quad\quad\quad\quad CH_3 \quad CH_3$$
$$\quad\quad\quad\quad\quad | \quad\quad | $$
$$CH_3 - CH_2 - C - CH - CH_3$$
$$\quad\quad\quad\quad\quad | $$
$$\quad\quad\quad\quad\quad CH_3$$

Q4 Name the following hydrocarbon:

$$CH_3CH_2CCCH(CH_3)CH_3$$

ANSWERS ⟫

1 2-methylpentane

2 but-1-ene

3 2,3,3-trimethylpentane

4 2-methylhex-3-yne

***Exam* tip:** In naming hydrocarbons, the basic name comes from the longest continuous chain. In Q1 this is five carbon atoms long so this is a pentane, with a methyl group on position 2. The carbon atoms are numbered from the end that gives the lowest numbers. That is why the compound in Q2 is but-1-ene and not but-3-ene.

The same rules apply in Q3. The longest continuous chain of carbon atoms is five carbon atoms so this is a pentane. The three methyl groups are numbered 2, 3, 3 from the right-hand end, because if we numbered the atoms from the left, they would be 3, 3, 4, which involves higher numbers.

In Q4, the bracketed group represents a methyl branch and the CC in the main chain tells us that this is an alkyne, with a carbon to carbon triple bond. Numbering the atoms from the right keeps the numbers as low as possible.

Substituted alkanes

Q1 In what way is the structure of an alkanone different from that of an alkanal?

Q2 Name the following compound:

$$CH_2OH$$
$$|$$
$$CHOH$$
$$|$$
$$CH_2OH$$

Q3 Which of the following compounds is an isomer of hexanal?

  A 2-methylbutanal   **B** 3-methylpentan-2-one
  C 2,2-dimethylbutan-1-ol   **D** 2-ethylpentanal

Q4 What is the shortened structural formula for pentanoic acid?

ANSWERS ▶▶

1 In an alkanone, the C=O (carbonyl group) is bonded to carbon atoms on either side. (It is in the 'middle' of the chain.) In an alkanal, the C=O group is at the end of the chain, and is bonded to a hydrogen atom on one side.

2 Propan-1,2,3-triol (glycerol).

3 **B**. 3-methylpentan-2-one has the same number of carbon, hydrogen and oxygen atoms as hexanal.

4 $CH_3CH_2CH_2CH_2COOH$

***Exam* tip:** In questions like Q3, in which you are asked to find an isomer of hexanal, it is worth remembering that alkanals and alkanones both have the general formula $C_nH_{2n}O$. This means that alkanals and alkanones with the same number of carbon atoms are isomers of each other. In Q3 the names indicate the number of carbon atoms. You don't need to draw out the structure.

A Five carbon atoms, so not an isomer of hexanal.

B A ketone with six carbon atoms, therefore an isomer of hexanal.

C An alcohol, so not likely to be an isomer of hexanal.

D Seven carbon atoms, so cannot be an isomer of hexanal.

Esters

Q1 Which of the following groups of atoms is the ester link?

A

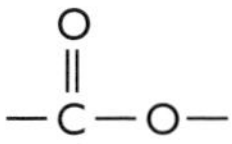

B

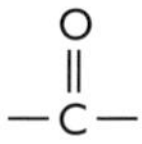

C

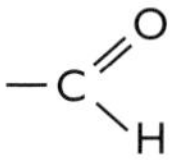

D

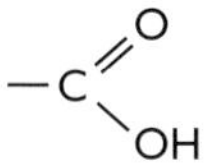

Q2 What would be the products of breakdown of the ester propyl butanoate?

Q3 Name the ester $C_2H_5COOC_2H_5$.

Q4 Here are the names of four esters. Which two are isomers of each other?

A pentyl ethanoate   **B** butyl methanoate
C propyl pentanoate   **D** butyl propanoate

ANSWERS ▶▶

1 **A**. In a shortened structural formula it may appear as —COO— .

   B is found in alkanones, **C** in alkanals and **D** in alkanoic acids.

2 Propanol and butanoic acid. The first part of the ester name comes from the alkanol used to form it. The second part of the name comes from the alkanoic acid involved.

3 The ester link is written —COO— in this structure. The C that has no H atoms bonded to it comes from the acid, so that the acid has three carbon atoms, propanoic acid. The alcohol must consist of the remaining two carbon atoms and be ethanol. Therefore, the name of this ester is ethyl propanoate.

4 **A**, **D**. If you draw out the structures of these molecules and count up all the atoms, you'll find that these are the isomers. It is useful to know that esters with the same numbers of carbon atoms are quite likely to be isomers of each other. If they don't have the same number of carbon atoms, they cannot be isomers of each other. The names tell us that the numbers of carbon atoms are:

   A 7 (pent + eth)   **B** 5 (but + meth)

   C 8 (prop + pent)   **D** 7 (but + prop)

***Exam* tip:** It can be helpful to note that esters and alkanoic acids with the same number of carbon atoms are also likely to be isomers of each other. For example, octanoic acid and pentyl propanoate are isomers. Oct = 8; pent + prop = 5 + 3 = 8.

Aromatic hydrocarbons

Q1 What is the molecular formula for benzene?

Q2 Benzene is extremely stable. What causes benzene to be so stable?

Q3 Here is the structure of an aromatic hydrocarbon:

a What is its molecular formula?
b Name this compound.

Q4 What term is used for this group of atoms?

ANSWERS

1 C_6H_6

2 Benzene is stable as a result of its delocalised electrons. You have already met delocalised electrons in metallic bonding, where they hold the lattice of positive metal ions together, and in graphite, where each sheet of C_6 rings is separated from the next by a layer of delocalised electrons.

3 a C_7H_8. Each corner of the hexagon represents a carbon. If nothing appears to be attached to a carbon atom it tells you that there is a hydrogen atom there. So there is C_6H_5 plus CH_3 giving C_7H_8.

  b methylbenzene or toluene

4 phenyl group

***Exam* tip:** Take care when you are asked for a molecular formula. In Answer **3a** the molecular formula is C_7H_8. If you had written $C_6H_5CH_3$ it would not, strictly speaking, have been a molecular formula but a shortened structural formula, and might not have earned full marks. To prevent you making an error of this type, the question might ask you to supply the values of x and y in the formula C_xH_y. In this case, $x = 7$ and $y = 8$.

Addition

Q1 Name three types of substance that can take part in addition reactions with **both** alkenes and alkynes to give saturated products.

Q2 Suggest how ethane, from natural gas, can be converted into ethanol (two steps.)

Q3 Propyne has the following structure:

$$CH_3 — C \equiv C — H$$

What are the shortened structural formulae for the **three** possible products when one mole of propyne reacts with two moles of hydrogen chloride (HCl)?

Q4 Which of the following statements about benzene is true?

A It is an isomer of cyclohexane.
B It reacts with bromine as if it is unsaturated.
C The carbon to hydrogen ratio is the same as in ethyne.
D It readily undergoes addition reactions.

ANSWERS ▶▶

1 hydrogen, halogens, hydrogen halides

2 Ethane must be cracked (heat + catalyst) to give ethene (it might be necessary to separate the ethene from other cracking products such as methane and hydrogen). Water must then be added (heat + catalysts) to give ethanol as the result of an addition reaction.

3 The products are $CH_3CCl_2CH_3$, $CH_3CH_2CHCl_2$ and $CH_3CHClCH_2Cl$.

   That is, both chlorine atoms end up on the middle carbon **or** both end up on the end carbon **or** one chlorine ends up on each. In fact, hydrogen atoms tend to add to carbon atoms that already are bonded to hydrogen atoms, so that the main product would be $CH_3CCl_2CH_3$.

4 **C**. Benzene is C_6H_6 and cyclohexane is C_6H_{12}. They are not isomers. Ethyne is C_2H_2 so the carbon to hydrogen ratio is the same.

***Exam* tip:** In Q4, options **B** and **D** are about the ability of benzene to take part in addition reactions. Benzene does not readily take part in such reactions due to the stability caused by its delocalised electrons.

Oxidation (1)

Q1 Classify the following alkanols as primary, secondary or tertiary:

   a $CH_3C(CH_3)(OH)CH_2CH_3$

   b $CH_3CH_2CH(OH)CH_2CH_3$

Q2 Name the two reagents commonly used to oxidise alkanols.

Q3 Insert the missing words: 'Primary alcohols are oxidised, first to _____ and then to _____ . Secondary alcohols are oxidised to _____ .'

Q4 Explain why the conversion of the compound $C_6H_4(OH)_2$ into $C_6H_4O_2$ can be regarded as oxidation.

ANSWERS ▶▶

1   **a** tertiary

    b secondary

    In **a**, the second C from the left carries an —OH group. The same C atom is bonded to three other C atoms.

    In **b**, the middle C atom carries an —OH group. The same C atom is bonded to two other C atoms, and a H atom.

2   hot copper oxide and acidified potassium dichromate solution

3   aldehydes (or alkanals); carboxylic (or alkanoic) acids; ketones (or alkanones)

4   The ratio of hydrogen to oxygen has fallen from 6:2 to 4:2.

***Exam* tip:** A simple way to judge whether an alcohol is primary, secondary or tertiary is as follows: if the —OH is at the *end* of a chain, it is primary; if it is in the *middle* of a chain it is secondary; if it is at a *branch point* (a carbon atom in the middle of the chain and carrying a branch, for example a methyl group) it is tertiary.

In Q4 we could also have said that the ratio of oxygen to hydrogen *rises* from 2:6 to 2:4.

Oxidation is taken to be an increase in the ratio of oxygen atoms to hydrogen atoms or a fall in the ratio of hydrogen atoms to oxygen atoms.

Oxidation (2)

Q1 When propan-2-ol is oxidised to propanone, the alcohol:
A loses 2 g per mole
B gains 2 g per mole
C gains 16 g per mole
D does not change mass

Q2 Propan-1-ol is made to react with hot copper oxide. The product is then treated with concentrated sulphuric acid and ethanol. Give the systematic name for the final product of these reactions.

Q3 Name an isomer of this compound:

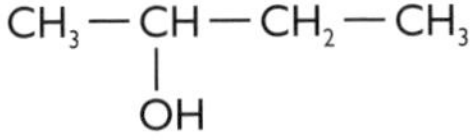

$$CH_3 - CH - CH_2 - CH_3$$
$$|$$
$$OH$$

(butan-2-ol)

that cannot be oxidised by hot copper oxide.

Q4 Name the compound that can be oxidised to give the compound with the structure shown below:

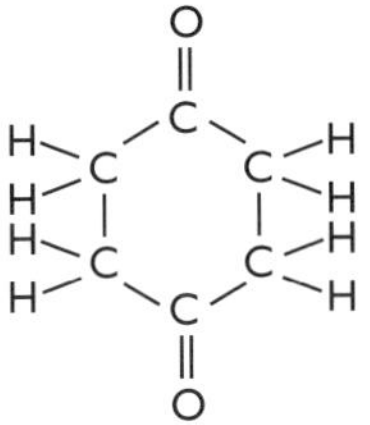

ANSWERS ▶▶

1   Oxidation of an alcohol to an aldehyde or ketone involves the loss of two hydrogen atoms. If one mole of aldehyde or ketone is oxidised, two moles of hydrogen atoms are lost, that is, 2 g. This is true no matter what aldehyde or ketone is involved.

2   As propan-1-ol is a primary alcohol, copper oxide will oxidise it first to propanal and then to propanoic acid. Treatment with ethanol and concentrated sulphuric acid gives an ester. Since this is an ester of propanoic acid and ethanol, it is ethyl propanoate.

3   Tertiary alcohols cannot be oxidised by hot copper oxide. The tertiary alcohol that is an isomer of butan-2-ol is 2-methylpropan-2-ol.

4   The starting compound must have hydroxyl groups (—OH) where the carbonyl groups have formed. The compound is cyclohexan-1, 4-diol.

Exam tip: It's important to remember that primary alcohols can be oxidised first to aldehydes, then to carboxylic acids. Secondary alcohols can be oxidised to ketones. Mild oxidising agents have no effect on tertiary alcohols. Tertiary alcohols are flammable, like other alcohols, which means that they can be oxidised to carbon dioxide and water.

Aldehydes and ketones

Q1 What is the main chemical difference between aldehydes and ketones?

Q2 Describe the Fehling's test, which distinguishes aldehydes from ketones.

Q3 **a** If Tollen's reagent is heated with an aldehyde, what is seen?
b If Tollen's reagent is heated with a ketone, what is seen?

Q4 Acidified potassium dichromate solution can be used to distinguish between aldehydes and ketones. What is seen when the solution is heated with an aldehyde?

ANSWERS ▶▶

1 Aldehydes are easily oxidised (to carboxylic acids) and ketones are not.

2 Fehling's solution (or Benedict's solution) is heated with the aldehyde or ketone in a warm water bath. An aldehyde turns the solution from blue to brown/red. A ketone has no effect.

3 a A silver mirror forms.
  b No change is seen.

4 The dichromate solution turns from orange to green/blue.

Exam tip: Fehling's or Benedict's solutions contain Cu^{2+} ions, which are blue. When they react with an aldehyde they oxidise the aldehyde to an acid. At the same time, they are reduced to Cu^+ ions, which accounts for the colour change. In Tollen's test, Ag^+ ions oxidise the aldehyde and are thus reduced to silver metal. In the dichromate test, the orange $Cr_2O_7^{2-}$ ion is reduced to the green/blue Cr^{3+} ion.

Making and breaking down esters (1)

Q1 The formation of an ester from an acid and an alcohol is a condensation reaction.

What is meant by a condensation reaction?

Q2 The breakdown of an ester into an acid and an alcohol is hydrolysis. What is meant by a hydrolysis reaction?

Q3 What catalyst is used in the formation of an ester from an acid and an alcohol?

Q4 An ester has the following structural formula:

$$CH_3 - CH_2 - \overset{\overset{\textstyle O}{\|}}{C} - O - \overset{\overset{\textstyle CH_3}{|}}{\underset{\underset{\textstyle CH_3}{|}}{\underset{|}{C}}} - H$$

Name the products obtained by hydrolysing this compound.

ANSWERS))

1 In a condensation reaction, two molecules combine and set free another small molecule. This small molecule is very often water, but it does not have to be water.

2 In a hydrolysis reaction a larger molecule is broken into smaller molecules by the addition of water. A hydrogen from water adds to one part of the larger molecule and a hydroxyl from water adds to the other. For example:

$$CH_3CH_2COOCH_3 + H_2O \longrightarrow CH_3CH_2COOH + HOCH_3$$

3 The catalyst often used is concentrated sulphuric acid.

4 The products are propanoic acid and butan-2-ol.

***Exam* tip:** In answering Higher Chemistry questions like Q3 and Q4 you must take care to give a complete answer. If you say that the catalyst is sulphuric acid (as opposed to *concentrated* sulphuric acid) you are likely to gain ½ mark rather than 1 mark, which is usually awarded for such a question.

If you say that the product is butanol (as opposed to *butan-2-ol*) you may also be penalised.

Making and breaking down esters (2)

Q1 When making an ester in the lab, a carboxylic acid and an alcohol are placed in a test tube with a few drops of concentrated sulphuric acid catalyst. The tube is heated in a warm water bath.

a Why is a wet paper towel wrapped round the tube near its mouth?

b Why is a plug of cotton wool placed in the mouth of the tube?

Q2 After heating, the contents of the tube are poured into a beaker containing a solution of sodium hydrogen carbonate.

How is the presence of an ester recognised?

Q3 What effect does the sodium hydrogen carbonate have on the catalyst?

Q4 Explain why unreacted reactants enter the solution of sodium hydrogen carbonate, while the ester does not.

ANSWERS ▶▶

1   a   The wet paper towel cools the tube, causing ester vapour to
        condense and run back into the tube.

    b   The cotton wool plug prevents hot reactants from spurting out of
        the tube.

2   The ester floats as a separate layer on the surface of the sodium
    hydrogen carbonate solution. It can also be recognised by its fruity or
    floral smell.

3   The sodium hydrogen carbonate neutralises the sulphuric acid catalyst.
    This means that once the ester has been formed, the catalyst cannot
    break the ester down again into carboxylic acid and alcohol.

4   Both the carboxylic acid and the alcohol are polar and therefore
    tend to dissolve in water. The ester is non-polar and therefore forms
    a separate layer. Being less dense than water the layer forms on the
    surface of the sodium hydrogen carbonate solution.

***Exam* tip:** The carboxylic acid and the alcohol both have hydroxyl
groups, —OH, which enable them to take part in hydrogen bonding with
water, resulting in solubility. Hydrogen bonding also gives them higher
boiling points than the ester, which has no hydroxyl groups. As a result
the ester with a lower boiling point is more likely to be lost from the
reaction mixture by evaporation. This is the reason for the use of the wet
paper towel.

Percentage yield

Q1 Percentage yield $= \dfrac{100 \times \underline{\hspace{1cm}} \text{ yield}}{\underline{\hspace{1cm}} \text{ yield}}$.

What are the two missing terms?

The next two questions refer to the following reaction:

$$CH_3OH + CH_3COOH \longrightarrow CH_3OOCCH_3 + H_2O$$

methanol   ethanoic acid   methylethanoate

Q2 If we start with 1.0 mole of methanol and 1.0 mole of ethanoic acid and end up with 0.25 mole of methylethanoate, what is the percentage yield?

Q3 If we start with 0.75 mole of methanol and 0.5 mole of ethanoic acid and end up with 0.25 mole of methylethanoate, what is the percentage yield?

Q4 Consider the following reaction:

$$C_2H_2 + 2Cl_2 \longrightarrow C_2H_2Cl_4$$

ethyne chlorine   1,1,2,2-tetrachloroethane

If we start with 1.0 mole of ethyne and 1.0 mole of chlorine and end up with 0.25 mole of tetrachloroethane, what is the percentage yield?

ANSWERS ▶▶

1   The missing words are 'actual' and 'theoretical'.

2   If we start with 1.0 mole of methanol and 1.0 mole of ethanoic acid
    the equation tells us that we should expect a maximum of 1.0 mole of
    methylethanoate. This is the theoretical yield. If we end up with only
    0.25 mole of product, then the percentage yield is

    $$\frac{100 \times 0.25}{1.0} \text{ or } 25\%.$$

3   If we start with 0.75 mole of methanol and 0.5 mole of ethanoic acid
    we can tell from the equation that we cannot expect more than
    0.5 mole of methylethanoate. It doesn't matter how much methanol
    we have; the theoretical yield is fixed by the reactant we have least of.
    This time the percentage yield is

    $$\frac{100 \times 0.25}{0.5} = 50\%.$$

4   In this example, we need 2.0 mole of chlorine to make 1.0 mole of
    tetrachloroethane. Therefore, if we only have 1.0 mole of chlorine, we
    can expect to make, at most, 0.5 mole of tetrachloroethane. In terms
    of the numbers of moles needed to make the product there is not
    enough chlorine. This time the percentage yield is

    $$\frac{100 \times 0.25}{0.5} = 50\%.$$

Exam tip: Often, in percentage yield calculations, quantities of
reactants and products are given as masses rather than in moles. It is
usually a good idea to convert given masses into moles and then to
compare the numbers of moles with the numbers of moles given in
the equation. Take care when an equation does not involve a simple
relationship of one mole of reactant to one mole of product!

Uses of carbon compounds

Q1 Which product is least likely to contain esters?

   A flavourings   **B** perfumes

   C solvents   **D** bleach

Q2 Give an example of the harm caused by the release of CFCs (chlorofluorocarbons) into the atmosphere.

Q3 Explain why we can describe fats and oils as esters.

Q4 Give a use for sodium stearate, and describe how it is made.

ANSWERS »

1 **D**. Bleach is least likely to contain esters. If you are asked for a use of esters, refer to perfumes, flavourings or solvents. Do not suggest any other application.

2 It is known that CFCs can react with ozone (O_3) in the upper atmosphere. Ozone has the ability to absorb ultra-violet radiation; if it is removed by reaction with CFCs, the radiation can penetrate to the surface of the earth where it can cause damage to plant and animal life.

3 When they are hydrolysed, fats and oils break down into glycerol, which is an alcohol, and various carboxylic acids such as stearic acid.

4 Sodium stearate is a soap. It is made by the hydrolysis of fats by sodium hydroxide. For example, glyceryl tristearate is hydrolysed to glycerol and stearic acid. The stearic acid then reacts with sodium hydroxide to form water and sodium stearate.

Exam tip: The Higher Chemistry course does not contain a great deal on the uses of carbon compounds. Most of the questions asked are about the uses of esters, or applications of halogenocarbons, including CFCs.

Early plastics and fibres

Q1 Ethene and propene are two important monomers. What is done to make the following?

  a ethene
  b propene

Q2 In what way does the monomer used to make an addition polymer differ from that used to make a condensation polymer?

Q3 What type of condensation polymer can be made from the following monomers?

  a $HO — (CH_2)_6 — OH$ and $HOOC — (CH_2)_4 — COOH$
  b $H_2N — (CH_2)_6 — NH_2$ and $HOOC — (CH_2)_4 — COOH$

Give a use for each of the above types of polymer.

Q4 Methanal is used in the manufacture of some thermosetting polymers. It is made from methanol, which is, in turn, made from synthesis gas.

  a Give the formula for methanol.
  b How is synthesis gas made?

ANSWERS 〉〉

1   **a** Ethene is made by cracking ethane from the gas fraction or by cracking naphtha from crude oil.

   b Propene is made by cracking propane from the gas fraction or by cracking naphtha from crude oil.

2   The monomer used for addition polymers must contain a carbon to carbon double bond; the monomer used for condensation polymers must have a functional group at each end.

3   **a** Polyester, used for fibres and resins.

   b Polyamide, an example of which is nylon, used for fibres and plastic engineering components.

4   **a** CH_3OH

   b Synthesis gas can be made by steam reforming of methane from natural gas, or by steam reforming of coal.

***Exam* tip:** When polyester *resins* are made, there must be additional bonding *between* the polyester chains. This is called 'cross-linking' and causes the polymer to become rigid. Polyamides contain amide links,

which allow hydrogen bonding to occur between the $C{=}O$ on one chain and the N—H on a neighbouring chain. This adds rigidity to the material.

Recent developments

Q1 Kevlar® is an aromatic polyamide with great strength in comparison to its weight. It is used for body armour and car tyres amongst other applications. Part of its strength results from intermolecular bonding. Name the type of intermolecular bonding involved and name the group of atoms responsible for the bonding.

Q2 What structural feature of polyethenol causes it to be soluble in water and why does it result in solubility?

Q3 Poly(ethyne) can be treated to make a polymer that conducts electricity.

What causes the polymer to be able to conduct electricity?

Q4 Low-density polythene can be modified during manufacture to form a photodegradable polymer.

What external factor will cause this polymer to degrade?

ANSWERS ▶▶

1 Hydrogen bonding is responsible; the group of atoms responsible is the amide link:

$$\begin{array}{cc} O & H \\ \| & | \\ -C & -N- \end{array}$$

which forms the link between all the monomer molecules.

2 The polyethenol molecule has —OH (hydroxyl) groups along its structure. As a result, it can take part in hydrogen bonding with water, and this assists solubility in water. The more —OH groups it has, the more soluble it is.

3 It conducts due to the presence of delocalised electrons.

4 light

Exam tip: Questions on recent develpments in polymers are often about the properties of the polymers. It is worth learning the following:

polymer	property
Kevlar®	strong
polyethenol	water soluble
polyethyne	electrical conductor
poly(vinylcarbazole)	photoconductor
biopol	biodegradable

You do not need to know the structures of the polymers.

Fats and oils

Q1 What are the three sources of fats and oils?

Q2 Fats and oils are esters.
  a Name the alcohol involved in all fats and oils.
  b Name the two main acids involved in most fats and oils.
  c What is the mole ratio between the acids and the alcohol?
  d In what way does the acid involved in oils differ from that in fats?

Q3 Oils are liquids at room temperature, because they have lower melting points than fats. What causes oils to have a lower melting point?

Q4 What role do fats and oils play in our diet?

ANSWERS ▶▶

1 animals, fish, plants

2 **a** glycerol

  b stearic acid and oleic acid

  c 3:1

  d In oils, the acid (often oleic acid) is unsaturated. In fats, the acids are saturated.

3 The presence of a double bond in the acid puts a bend into the structure of the acid molecule and therefore into the oil molecule. This means that the molecules are unable to pack together as closely as they do in fats, where there is no bend. As a result, the oil is liquid.

4 They supply the body with energy in a more concentrated form than carbohydrates do.

Exam tip: Although stearic and oleic acids are the most frequently occurring acids involved in fats and oils, the acids can range from C_4 to C_{24}, but C_{16} and C_{18} are the most common. The acids always have an even number of carbon atoms. Stearic acid is $CH_3(CH_2)_{16}COOH$ and oleic acid is $CH_3(CH_2)_7CH=CH(CH_2)_7COOH$ (the double bond is in the middle of the molecule). It is worth learning these two formulae.

Proteins (1)

Q1 Name the four elements found in every protein.

Q2 Proteins are built from large numbers of amino acids linked together.
  a What *two* functional groups are found in every amino acid?
  b Name the 'link' formed each time two amino acids combine.
  c Each time a link forms, a water molecule is set free.
    What term is used to describe this type of reaction?

Q3 The body can make most of the amino acids required for protein formation.
  a How does it obtain the amino acids it cannot make?
  b What term is used to describe such amino acids?

Q4 The breakdown (digestion) of protein molecules in food is hydrolysis.
  What happens in a hydrolysis reaction?

ANSWERS))

1 carbon, hydrogen, oxygen, nitrogen (you can remember this easily if you remember CHON)

2 a Each amino acid molecule contains an amino group, —NH_2, and a carboxyl group, —COOH.
   b The link is an amide or peptide link.
   c condensation reaction

3 a These amino acids are taken in from the diet.
   b They are called 'essential' amino acids.

4 A longer molecule is broken down into shorter molecules by the addition of hydrogen and oxygen from water.

***Exam* tip:** Amino acids all have the basic structure shown below.

R stands for various groups of atoms. If R = H, the amino acid is glycine. If R = CH_3 it is alanine. These are not names you have to remember. When amino acids combine in a condensation reaction, a peptide or amide link forms at each join. The link is as shown below:

$$\begin{array}{cc} O & H \\ \| & | \\ -C-N- \end{array}$$

Proteins (2)

Q1 Proteins do not exist simply as long chains of amino acids but are folded or coiled into more complex structures. Bearing in mind that there are peptide links all along the chain, suggest what intramolecular forces hold the protein in its complex structure.

Q2 Proteins can be classified as 'fibrous' or 'globular'.

  a How does the shape of a globular protein differ from that of a fibrous protein?

  b How does the body use fibrous proteins?

  c How does the body use globular proteins?

Q3 Enzymes are highly specific, that is they operate on one molecular structure only, the substrate. What is thought to make enzymes so specific?

Q4 **a** What factors can cause an enzyme to become denatured?

  b What happens to the enzyme molecule when it is denatured?

ANSWERS))

1 The peptide link has the structure:

Hydrogen bonding can occur between the oxygen atom of one peptide link and the hydrogen atom of another, holding the molecule in a more complex shape.

2 a Fibrous protein molecules are long and thin. Globular proteins are shorter and rounder.

   b Fibrous proteins form long, thin, strong bundles and usually have a structural role. Fibrous proteins are found in skin, hair and muscle.

   c Globular proteins are involved in the maintenance and regulation of life processes. Examples of globular proteins include haemoglobin (responsible for carrying oxygen around the body), hormones, such as insulin, and all enzymes.

3 The surface of an enzyme is folded in a particular way, so that only one shape of molecule can fit into it. This explanation is often called the 'lock and key' model.

4 a Excessively high temperatures and extreme pH values can result in denaturation.

   b Denaturation involves a structural change to the enzyme so that the shape of its surface is altered and it can no longer accommodate its substrate molecule.

***Exam* tip:** Saliva contains an enzyme called amylase. This enzyme breaks starch down into sugars, which is why, if you chew a piece of bread long enough, it begins to taste sweet.

Industry makes use of enzyme-catalysed reactions, partly because they are highly specific (they act only on one substrate) and they require very little energy to proceed.

Proteins (3)

Q1 The following apparatus is used to investigate the effect of pH on the activity of an enzyme that breaks down hydrogen peroxide, H_2O_2, in our bodies.

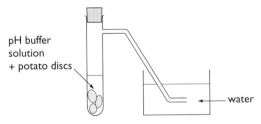

pH buffer solution + potato discs

water

Name the enzyme involved.

Q2 After the above apparatus has been left for three minutes, hydrogen peroxide is added. What are the products of the breakdown of hydrogen peroxide?

Q3 **a** How is the activity of the enzyme measured?
b Special pH buffer solutions can be used to give pH values of 4, 7 and 9.
What solutions are used to give the following pH values?

   i 1
   ii 13

Q4 Explain why the activity of the enzyme decreases as very high or low pH values are approached.

ANSWERS

1 catalase

2 The products of the breakdown are water and oxygen.

The equation for the reaction is:

$$H_2O_2 \longrightarrow H_2O + \tfrac{1}{2}O_2.$$

3 a The reaction produces oxygen gas, which passes through the delivery tube and bubbles through the water in the beaker. The rate of reaction at different pH values is found by comparing the number of bubbles formed per minute for the first three minutes of reaction.

  b i $0.1 \, mol \, l^{-1}$ hydrochloric acid

    ii $0.1 \, mol \, l^{-1}$ sodium hydroxide

4 At high or low pH values, the protein molecules of the enzyme become denatured.

Exam tip: This PPA can also be used to investigate the effect of temperature on the activity of the enzyme. In this case, deionised water is used instead of buffer solution and the side-arm test tube is placed in a warm water bath until the contents of the tube have reached a steady temperature. Hydrogen peroxide is then added and the rate of oxygen production measured as described above.

The chemical industry (1)

Q1 Chemical industrial processes can be described as 'batch' or 'continuous'.

a Give two advantages of a batch process.
b Give two advantages of a continuous process.

Q2 What is meant by the term 'raw material'?

Q3 Which of the following compounds is a raw material in the chemical industry?

A propene   **B** ammonia
C sulphuric acid   **D** sodium chloride

Q4 What is meant by the term 'feedstock'?

ANSWERS

49 The chemical industry (1)

1 a Batch processes are suitable for preparing small quantities of product and the same chemical plant can be used to make different products.

   b Continuous processes are suitable for making large quantities of product and require minimal labour.

2 Raw materials are substances that are obtained from the earth's crust, and used without undergoing any major change to prepare them for use.

3 D. Sodium chloride is the only substance in the list that is found in the Earth's crust, often in underground deposits. The three other substances in the list must all be made by a series of chemical reactions.

4 A feedstock is a chemical that is used in an industrial process to make other chemicals. It is often made from 'raw materials'.

Exam tip: Raw materials include air, water, sea water, crude oil, natural gas, coal, untreated metal ores and sand. Natural gas is a raw material, but pure methane would be considered a feedstock as it is separated from liquid natural gas by fractional distillation. Air is a raw material, but oxygen is a feedstock as it has to be separated from air, by fractional distillation of liquid air.

The chemical industry (2)

Q1 What is the main capital cost in any industrial chemical process?

Q2 Give an example of a variable cost in an industrial chemical process.

Q3 Give an example of a factor that must be taken into account when deciding on the location of a new chemical plant.

Q4 Give an example of how the chemical industry attempts to minimise damage to the environment.

ANSWERS

1   The main capital cost is the cost of building the chemical plant itself.

2   Variable costs include the costs of raw materials, the costs of waste treatment and disposal, and the costs of distributing the product.

3   There should be good transport links, so that raw materials can easily reach the plant and product can be easily distributed. If the plant is near a population centre, there is a potential workforce available. It is also useful if the plant is near other plants that produce some of the other materials required.

4   The chemical industry tries to minimise the release of gases into the air. It treats liquid waste before releasing it into rivers or the sea. It also recycles materials where possible.

Exam tip: The cost of raw materials is a *variable*, since, if the plant is not working at full capacity, it uses less raw materials, produces less waste and there is less product to distribute (by road, rail, etc.). Another type of cost is a *fixed cost*. These are costs incurred by the company no matter how much or little product it makes. They include labour costs, rates or council tax and the cost of depreciation of the plant. (As the plant gets older, it is worth less, whether or not it is making product.)

Hess's Law

Q1 What is meant by 'Hess's Law'?

Q2 Use the following data to calculate ΔH for the formation of sulphur trioxide from sulphur and oxygen.

$S + O_2 \longrightarrow SO_2$ $\Delta H = -298 \text{ kJ mol}^{-1}$

$SO_2 + \frac{1}{2}O_2 \longrightarrow SO_3$ $\Delta H = -99 \text{ kJ mol}^{-1}$

Q3 Here is information on two combustion reactions:

$CO + \frac{1}{2}O_2 \longrightarrow CO_2$ $\Delta H = -283 \text{ kJ mol}^{-1}$

$Cu + \frac{1}{2}O_2 \longrightarrow CuO$ $\Delta H = -155 \text{ kJ mol}^{-1}$

Use this information to find the enthalpy change for the reaction:

$CuO + CO \longrightarrow Cu + CO_2$

Q4 Given that the enthalpy of neutralisation for potassium hydroxide *solution* with hydrochloric acid is -58 kJ mol^{-1} and the enthalpy of solution of *solid* potassium hydroxide is -55 kJ mol^{-1}, use Hess's Law to calculate the enthalpy change for the reaction:

$KOH(s) + HCl(aq) \longrightarrow KCl(aq) + H_2O(l)$

ANSWERS ▶▶

1 Hess's Law states that the energy change as reactants turn into products is independent of the route taken. For example, we can burn a mole of carbon to form a mole of carbon dioxide releasing X kJ of energy. If we convert the carbon first into carbon monoxide and then burn the carbon monoxide to form carbon dioxide the overall energy change will still be X kJ.

2 By cancelling the formula for SO_2, which appears on both sides of the equations, and adding the remaining terms we find that $\Delta H = -397$ kJ mol^{-1}.

3 By reversing the second equation *and the sign of its ΔH,* the $\frac{1}{2}O_2$ terms will cancel, leaving the required formulae on the correct sides of the equation. Adding gives the required equation and the value $\Delta H = -128$ kJ mol^{-1}.

4 Since the equation involves the KOH first dissolving, and then neutralising the HCl, ΔH for this reaction is the sum of the enthalpies of solution and neutralisation, giving $\Delta H = -113$ kJ mol^{-1}.

***Exam* tip:** If we add equations representing chemical reactions, we get an equation that represents a new reaction. If we add the ΔH values in the same way, we get ΔH for the new reaction. If we multiply an equation, the value of ΔH has to be multiplied in the same way. If we reverse an equation, the value of ΔH has to be reversed as well, that is, its sign must be reversed.

Dynamic equilibrium

Q1 How does the rate of the forward reaction compare with the rate of the reverse reaction:

   a before equilibrium is reached?
   b after equilibrium is reached?

Q2 How does the concentration of:

   a reactant
   b product

change before equilibrium is reached?

Q3 Explain why it is wrong to say that, at equilibrium, reactants cease to turn into products and products cease to turn into reactants.

Q4 What can we say about the concentrations of reactants and products at equilibrium?

ANSWERS

1 **a** Before equilibrium is reached, the forward reaction is faster than the reverse reaction. (If there is no product present at the start, then the rate of the reverse reaction must be zero at the start.)

b Once equilibrium is reached, forward and reverse reactions have the same rate.

2 **a** Before equilibrium is reached, the reactant concentration falls.

b Before equilibrium is reached, the product concentration rises.

3 At equilibrium reactants are still turning into products, and products into reactants, but they do so at the same rate so there is no overall change in concentration.

4 At equilibrium the concentrations of reactants and products remain constant.

Exam tip: 'Dynamic' equilibrium means that, although there is no *overall* conversion of reactant into product or of product into reactant at equilibrium, reactants are steadily turning into products and products into reactants, *but at the same rate*.

At equilibrium, therefore, the concentrations of reactants and products remain constant, but they are not necessarily equal. In fact, they are almost always unequal.

Shifting the equilibrium position (1)

Q1 Consider the reaction A + B $\rightleftharpoons$ C + D ΔH negative.

Explain what effect an increase in temperature will have on the position of equilibrium.

Q2 Consider the reaction 2A(g) + B(g) $\rightleftharpoons$ C(g) + D(g).

Explain what effect an increase in pressure will have on the position of equilibrium.

Q3 Consider the reaction A + B $\rightleftharpoons$ C + D.

Explain what effect an increase in the concentration of D will have on the position of equilibrium.

Q4 Consider the reaction A + B $\rightleftharpoons$ C + D.

Explain what effect the presence of a catalyst will have on the position of equilibrium.

ANSWERS ▶▶

1 Since ΔH is negative, the forward reaction is exothermic. This means that the reverse reaction is endothermic – it takes in heat. This is favoured by a rise in temperature so the reverse reaction is favoured and the equilibrium moves to the left.

2 There are three moles of gas on the left of the equation and two on the right.

   The volume ratio of reactants to products is 3:2 as gas volumes are proportional to the number of moles. A pressure increase favours the smaller volume, so the equilibrium moves to the right.

3 Increasing the concentration of D increases the number of collisions between C and D. This favours the reverse reaction so the equilibrium moves to the left.

4 The presence of a catalyst does not affect the position of equilibrium.

***Exam* tip:** If we constantly remove a product as it forms, the reverse reaction cannot take place at all. This drives the equilibrium steadily to the right (the product side) until conversion is complete.

A catalyst lowers the activation energy of the forward and the reverse reactions equally. It causes equilibrium to be reached *faster*, but does not affect the *position* of equilibrium.

Shifting the equilibrium position (2)

When chlorine is added to water, the following equilibrium is set up:

$$Cl_2(g) + H_2O(l) \rightleftharpoons Cl^-(aq) + ClO^-(aq) + 2H^+(aq)$$

Q1 What effect would the addition of more chlorine have on the position of equilibrium?

Q2 What effect would the addition of solid NaCl have on the position of equilibrium?

Q3 What would be the effect on the equilibrium position of adding

a acid
b alkali

to the mixture?

Q4 How would an increase in pressure affect the position of equilibrium?

ANSWERS ▶▶

1   An increase in the chlorine concentration would increase the collision rate between Cl_2 and H_2O, resulting in an increase in the rate of the forward reaction. The equilibrium would move to the right.

2   The addition of solid NaCl would add Cl^- ions to the mixture. This would increase the collision rate between Cl^- and other species on the right-hand side of the equilibrium, resulting in an increase in the rate of the reverse reaction.

    The equilibrium would move to the left.

3   a   The addition of acid would add H^+ ions to the mixture. This would increase the collision rate between H^+ and the other species on the right-hand side of the equilibrium, resulting in an increase in the rate of the reverse reaction.

        The equilibrium would move to the left.

    b   The addition of alkali would remove H^+ from the mixture. This would reduce the collision rate between H^+ and the other species on the right. This would lower the rate of the reverse reaction.

        The equilibrium would move to the right.

4   There is one mole of gas on the left-hand side and none on the right. An increase in pressure causes the equilibrium to move to the side with the smaller volume.

    The equilibrium would move to the right.

Exam tip: In general, adding reactants to an equilibrium mixture moves the equilibrium to the right and adding products moves it to the left. Removal of products slows the reverse reaction and encourages the formation of products.

Shifting the equilibrium position (3)

Q1 The equation for the Haber process is:

$$N_2(g) + 3H_2(g) \rightleftharpoons 2NH_3(g) \qquad \Delta H = -92\,kJ\,mol^{-1}$$

Explain why the process is carried out at about 500 °C.

Q2 Explain why the reaction is carried out at a moderately high pressure.

Q3 Name the catalyst used in the process.

Q4 What two features of the process (apart from any of the factors mentioned in Q1–Q3) ensure that there is a high conversion of reactants into products?

ANSWERS ⟩⟩

1 Although this reaction is exothermic and is favoured by lower temperatures the reaction would be too slow at lower temperatures. The temperature used is a compromise between equilibrium (favoured by lower temperatures) and rate (favoured by higher temperatures).

2 This reaction involves four moles of gas becoming two moles, that is four volumes into two volumes. A high pressure favours the smaller volume. However, since high pressures are costly to maintain and can be hazardous, the pressure used is again a compromise between equilibrium (favoured by higher pressure) and costs/safety (favoured by lower pressure).

3 iron

4 The ammonia is chilled and liquefied, and so removed from the gas mixture, while unreacted nitrogen and hydrogen are recycled into the system. Removing ammonia as it forms prevents the reverse reaction (breakdown of ammonia) from taking place.

Exam tip: The catalyst used is iron, but oxides of some other metals are added as 'promoters' to enhance catalyst action. The iron is present as small pieces to give it a large surface area. Since iron is not expensive, it can be replaced from time to time; it does not have to be regenerated.

The pH scale

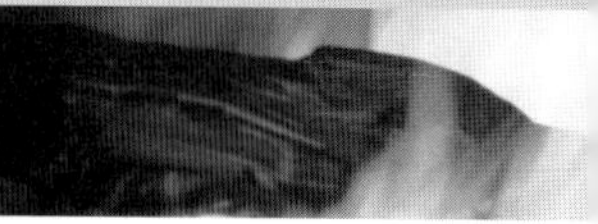

Q1 What is the hydrogen ion concentration, in mol l^{-1}, in solutions with the following pH values?

a 3
b 5
c 9

Q2 State the relationship between the concentration of H$^+$ ions and the concentration of OH$^-$ ions in an aqueous solution.

Q3 What is the hydroxide ion concentration, in mol l^{-1}, in solutions with the following pH values?

a 3
b 5
c 9

Q4 In water the following equilibrium exists:

$$H_2O(l) \rightleftharpoons H^+(aq) + OH^-(aq)$$

a What is the concentration of each ion in pure water?
b Does the equilibrium position favour the reactant side or the product side?

ANSWERS ▶▶

1 a 10^{-3}

  b 10^{-5}

  c 10^{-9}

2 In all aqueous solutions, the concentration of H^+ ions, in $mol\,l^{-1}$, multiplied by the concentration of OH^- ions, in $mol\,l^{-1}$, is 10^{-14}. That is, $[H^+][OH^-] = 10^{-14}$.

3 Since $[H^+][OH^-] = 10^{-14}$, if we know the concentration of H^+ ions, we can work out the concentration of OH^- ions by dividing 10^{-14} by the concentration of H^+ ions.

So the answers are:

a $\dfrac{10^{-14}}{10^{-3}} = 10^{-11}$

b $\dfrac{10^{-14}}{10^{-5}} = 10^{-9}$

c $\dfrac{10^{-14}}{10^{-9}} = 10^{-5}$

4 a In pure water and in neutral aqueous solutions, the concentration of each ion is $10^{-7}\,mol\,l^{-1}$. $[H^+] = [OH^-] = 10^{-7}\,mol\,l^{-1}$.

  b The equilibrium lies well to the reactant side. That is, nearly all the water is present as molecules and there are very few ions.

***Exam* tip:** The pH of any solution can be calculated from the relationship:

$$pH = -\log_{10}[H^+]$$

Values below pH 7 correspond to an acidic solution, while those above seven correspond to alkaline solutions. When pH = 7 a solution is neutral. Evidence for Answer 4b is shown by the fact that pure water has a very low electrical conductivity, so very few ions can be present.

The concept of strong and weak (1)

Q1 Hydrochloric acid, HCl, is an example of a strong acid, while ethanoic acid, CH_3COOH, is an example of a weak acid.

Describe the difference between strong and weak acids.

Q2 Considering $0.1\,mol\,l^{-1}$ solutions of hydrochloric acid and ethanoic acid:

a How does the pH of hydrochloric acid compare with that of ethanoic acid?

b How does the electrical conductivity of hydrochloric acid compare with that of ethanoic acid?

c How does the rate of reaction between hydrochloric acid and magnesium compare with the rate of reaction between ethanoic acid and magnesium?

Q3 Explain why even though one mole of ethanoic acid contains very few H^+ ions, it can still neutralise one mole of sodium hydroxide.

Q4 Name the weak acids formed by the solution of the following gases in water:

a carbon dioxide

b sulphur dioxide.

ANSWERS ▶▶

1 Strong acids, like HCl, break down (dissociate) completely into ions. Weak acids, like CH_3COOH, break down (dissociate) only partially into ions. Most of the weak acid remains as molecules.

2 a The pH of HCl is lower than that of CH_3COOH.

  b The conductivity of HCl is higher than that of CH_3COOH.

  c Magnesium reacts more quickly with HCl than with CH_3COOH.

This all follows from the fact that HCl contains a greater concentration of H^+ ions than does CH_3COOH.

3 Ethanoic acid takes part in the following equilibrium:

$$CH_3COOH(aq) \rightleftharpoons CH_3COO^-(aq) + H^+(aq)$$

When OH^- ions are added to the acid, they combine with and remove H^+ ions. More of the acid breaks down to replace them. In turn, the new H^+ are removed and this continues until all the acid has broken down (dissociated) and all the H^+ ions are removed. The equilibrium keeps shifting to the right.

4 a carbonic acid

  b sulphurous acid

***Exam* tip**: If we represent acids, in general, as HA, then they take part in the equilibrium $HA \rightleftharpoons H^+ + A^-$.

In a strong acid, this equilibrium lies well to the right. The products (ions) are favoured. In a weak acid, the equilibrium lies well to the left. The reactants (molecules) are favoured. We can say that strong acids exist mainly as ions, while weak acids exist mainly as molecules.

The concept of strong and weak (2)

Q1 Sodium hydroxide, NaOH, is an example of a strong base, while ammonium hydroxide, NH_4OH, is an example of a weak base.

Describe the difference between strong and weak bases.

Q2 Considering 0.1 mol l^{-1} solutions of sodium hydroxide and ammonium hydroxide:

a How does the pH of sodium hydroxide compare with that of ammonium hydroxide?

b How does the electrical conductivity of sodium hydroxide compare with that of ammonium hydroxide?

Q3 Explain why even though one mole of ammonium hydroxide contains very few OH$^-$ ions, it can still neutralise one mole of hydrochloric acid.

Q4 Ammonium hydroxide can be made by dissolving the gas ammonia, NH_3, in water. Using an equation, explain why this results in a solution of ammonium hydroxide.

ANSWERS ▶▶

1 Strong bases, like NaOH, break down (dissociate) completely into ions. Weak bases, like NH_4OH, break down (dissociate) only partially into ions. Most of the weak base remains as molecules.

2 a The pH of the NaOH solution is higher than that of the NH_4OH solution.

  b The conductivity of the NaOH solution is higher than that of the NH_4OH solution.

Both results arise from the fact that the NaOH solution has a greater concentration of OH^- ions than the NH_4OH solution. Like H^+ ions, OH^- ions are good conductors of electricity.

3 Ammonium hydroxide takes part in the following equilibrium:

$$NH_4OH(aq) \rightleftharpoons NH_4^+(aq) + OH^-(aq)$$

When H^+ ions are added to the base, they combine with and remove OH^- ions. More of the base breaks down to replace them. In turn, the new OH^- ions are removed and this continues until all the base has broken down (dissociated) and all the OH^- ions are removed. The equilibrium keeps shifting to the right.

4 $NH_3(g) + H_2O(l) \rightleftharpoons NH_4OH(aq) \rightleftharpoons NH_4^+(aq) + OH^-(aq)$

Since NH_4OH is a weak base, there are few ions present. Most exists as NH_4OH molecules or NH_3 molecules in water.

***Exam* tip:** If we represent bases, in general, as BOH, then they take part in the equilibrium $BOH \rightleftharpoons B^+ + OH^-$.

In a strong base, this equilibrium lies well to the right. The products (ions) are favoured. In a weak base, the equilibrium lies well to the left. The reactants (molecules) are favoured. We can say that strong bases exist mainly as ions, while weak bases exist mainly as molecules.

pH of salt solutions

Q1 Explain why a solution of sodium ethanoate is alkaline.

Q2 Explain why a solution of ammonium chloride is acidic.

Q3 Explain why a solution of sodium chloride is neutral.

Q4 A solution of potassium cyanide in water is alkaline. What does this tell us about the acid hydrogen cyanide?

ANSWERS ▶▶

1 The salts of weak acids and strong alkalis are alkaline. Sodium ethanoate, $Na^+CH_3COO^-$, is the salt of the weak acid, ethanoic acid, and the strong alkali, sodium hydroxide.

   In water, the ions Na^+ and CH_3COO^- separate from each other.

   The ethanoate ion tries to become stable by forming ethanoic acid so it reacts with water:

   $$CH_3COO^- + H_2O \longrightarrow CH_3COOH + OH^-$$

   The hydroxide ions produced make the solution alkaline.

2 The salts of weak alkalis and strong acids are acidic. Ammonium chloride, $NH_4^+Cl^-$, is the salt of the weak alkali, ammonium hydroxide, and the strong acid, hydrochloric acid.

   In water, the ions NH_4^+ and Cl^- separate from each other.

   The ammonium ion tries to become stable by forming ammonium hydroxide so it reacts with water:

   $$NH_4^+ + H_2O \longrightarrow NH_4OH + H^+$$

   The hydrogen ions produced make the solution acidic.

3 Sodium chloride is the salt of a strong acid, hydrochloric acid, and a strong base, sodium hydroxide. When added to water, the ions Na^+ and Cl^- separate from each other. Since strong acids and bases tend to exist as ions rather than as molecules, there is no reaction with water molecules to form a surplus of either H^+ or OH^- ions.

4 Since salts of weak acids and strong bases are alkaline, hydrogen cyanide must be a weak acid.

Exam **tip:** A simple way to remember how salts of weak acids and bases behave is to remember that 'the strong one wins'. If we have a salt of a weak acid and a strong base, the solution is alkaline. If the salt is from a strong acid and a weak base, then it is acidic.

Oxidising and reducing agents (1)

Q1 Explain what is meant by:

  a oxidation
  b reduction.

Q2 Explain what is meant by the terms:

  a oxidising agent
  b reducing agent.

Q3 In the reaction:

$Zn + 2HCl \longrightarrow ZnCl_2 + H_2$

identify:

  a the species that has been oxidised
  b the species that has been reduced
  c the oxidising agent
  d the reducing agent.

Q4 Which ion is oxidised in the following reaction?

$HgCl_2 + SnCl_2 \longrightarrow Hg + SnCl_4$

ANSWERS ▶▶

1 **a** loss of electrons

    b gain of electrons

2 **a** An oxidising agent removes electrons from another chemical.

    b A reducing agent supplies electrons to another chemical.

3 **a** The Zn has been oxidised, since it has lost electrons to form Zn^{2+}.

    b The H^+ ions have been reduced, since they have gained electrons to become H_2.

    c The oxidising agent is H^+, since it has removed electrons from Zn.

    d The reducing agent is Zn, since it has supplied electrons to H^+ ions.

4 In $SnCl_2$ the tin is present as Sn^{2+}. In $SnCl_4$, it is present as Sn^{4+}. This means that Sn in $SnCl_2$ has lost two electrons, so it has been oxidised.

***Exam* tip:** You can remember what oxidation and reduction are if you remember 'OILRIG'; Oxidation Is Loss, Reduction Is Gain' (of electrons). Redox reactions occur when both reactions happen together, as they usually do.

An *agent* causes something to happen to something else. So an oxidising agent oxidises something else, while itself becoming reduced.

Oxidising and reducing agents (2)

Q1 Where, in the electrochemical series (ECS), are

a   the most powerful oxidising agents
b   the most powerful reducing agents?

Q2 In which of the following reactions is hydrogen acting as an oxidising agent?

A $H_2 + C_2H_4 \longrightarrow C_2H_6$
B $H_2 + Cl_2 \longrightarrow 2HCl$
C $H_2 + 2Na \longrightarrow 2NaH$
D $H_2 + CuO \longrightarrow H_2O + Cu$

Q3 Complete the following ion–electron equation for the oxidation:

$CH_3CHO \longrightarrow CH_3COOH$.

Q4 Look at the following equation:

$2Ag^+ + C_6H_6O_2 \longrightarrow C_6H_4O_2 + 2Ag + 2H^+$

What is the ion–electron equation for the oxidation reaction?

ANSWERS ▶▶

61 Oxidising and reducing agents (2)

1 a The most powerful oxidising agents are at the foot of the ECS.

   b The most powerful reducing agents are at the top of the ECS.

2 Hydrogen usually acts as a reducing agent. However, look at equation **C**. Na metal ends up as the ionic compound NaH, which contains positive sodium ions and the *negative* hydride ion, H^-. The hydrogen has *taken* electrons from the sodium, forming Na^+ ions. The hydrogen is an oxidising agent in this reaction.

3 There is one more oxygen on the right-hand side than on the left. We add a water molecule to the left-hand side to give:

$CH_3CHO + H_2O \longrightarrow CH_3COOH$

We now add two hydrogen ions to the right-hand side to give:

$CH_3CHO + H_2O \longrightarrow CH_3COOH + 2H^+$

Finally, we need two electrons on the right to balance the charge:

$CH_3CHO + H_2O \longrightarrow CH_3COOH + 2H^+ + 2e^-$

4 The reduction reaction is $2Ag^+ \longrightarrow 2Ag$ (silver ions have to gain electrons for this to take place).

Removing this from the given equation leaves:

$C_6H_6O_2 \longrightarrow C_6H_4O_2 + 2H^+$.

Everything here is balanced, except the charge. We balance that by adding two electrons to the right-hand side, giving:

$C_6H_6O_2 \longrightarrow C_6H_4O_2 + 2H^+ + 2e^-$.

***Exam* tip:** The standard routine for producing balanced ion–electron equations is as follows:

1 Balance all elements except hydrogen and oxygen.
2 Balance the oxygen by adding water molecules as required to the side short of oxygen.
3 Balance the hydrogen by adding hydrogen ions to the side short of hydrogen.
4 Balance the charge using electrons.

Redox titrations

Q1 The amount of vitamin C in a tablet can be found by titration with iodine solution. The ion–electron equations for the reactions involved are:

$$C_6H_8O_6 \text{ (vitamin C)} \longrightarrow C_6H_6O_6 + 2H^+ + 2e^-$$

$$I_2 + 2e^- \longrightarrow 2I^-.$$

For each mole of iodine used, how many moles of vitamin C are present?

Q2 The iodine is added from a burette to the solution of vitamin C. What indicator is used to show when all the vitamin C has reacted?

Q3 How is the end point recognised?

Q4 One mole of vitamin C weighs 176 g. A solution of vitamin C contains 0.176 g of vitamin C.

   a How many moles of vitamin C are present in the solution?

   b What volume of a $0.02 \, mol \, l^{-1}$ iodine solution would be required to react with this amount of vitamin C?

ANSWERS ▸▸

1 The equations tell us that one mole of vitamin C loses two moles of electrons and that one mole of iodine accepts two moles of electrons. It follows that one mole of vitamin C reacts with one mole of iodine.

2 Starch solution is added to the vitamin C solution in the flask below the burette.

3 Before any iodine is added, the solution should be colourless, although many vitamin C tablets contain colouring, which results in the solution being coloured. As iodine is added, it is reduced to iodide ions by the vitamin C, which in turn is oxidised. Once all the vitamin C has been oxidised, the iodine is no longer reduced to iodide and so free iodine is present in the solution, causing a blue–black colour with the starch.

4 A mass of 0.176 g of vitamin C is $\frac{0.176}{176} = 0.001$ mole.

The amount of iodine required is therefore 0.001 mole.

If the iodine is 0.02 mol l^{-1}, the volume required is:

$\frac{0.001}{0.02} = 0.05$ litre (50 ml).

***Exam* tip:** When giving the colour change in a reaction it is necessary to give both the starting colour and the finishing colour. The correct answer to Q3 is 'colourless to blue–black'.

Electrolysis (1)

Q1 Calculate the number of coulombs passed through a circuit using a current of:

   a 1 A for 30 seconds
   b 5 A for 2 minutes
   c 10 A for 1 minute.

Q2 **a** How many electrons are there in one mole of electrons?
   b How many coulombs are carried by one mole of electrons?

Q3 What term is used for the amount of charge carried by one mole of electrons?

Q4 How many moles of electrons are required to reduce the following?

   a 1 mole of Na^+ to Na metal
   b 1 mole of Mg^{2+} to Mg metal
   c Al^{3+} ions to 27 g of Al metal

ANSWERS ▶▶

1 The number of coulombs is found by multiplying amps by the number of seconds.

This gives:

a 30 C

b 600 C

c 600 C

2 a 6.02×10^{23} electrons (the Avogadro number)

   b 96 500 C

3 A faraday, symbol F (= 96 500 C).

4 a 1 mole

   b 2 moles

   c 3 moles

***Exam* tip:** In the reduction of metal ions to metal, the general equation is

$$M^{n+} + ne^- \longrightarrow M.$$

To find the current or time required to form a given mass of metal from its ions:

a Convert the mass to moles by dividing by the RAM of the metal.

b Use this to calculate the number of moles of electrons required (faradays).

c Multiply the number of faradays by 96 500 to find the number of coulombs.

d Divide the number of coulombs by the current to find the number of seconds required, or by the number of seconds to find the current required.

Electrolysis (2)

Q1 To find the number of coulombs needed to produce one mole of hydrogen gas, the following apparatus is used.

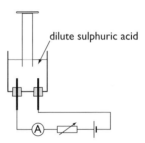

dilute sulphuric acid

a What is the ion–electron equation for the formation of hydrogen using this apparatus?

b How many coulombs are required to form one mole of hydrogen?

Q2 Over which electrode (left or right) should the measuring cylinder be placed to collect hydrogen?

Q3 In an experiment using this apparatus, $24\,cm^3$ of hydrogen was collected.

Given that the molar volume of hydrogen is 24 litres, calculate the number of coulombs required to form this volume of hydrogen.

Q4 The circuit contains a variable resistor. How can this improve the accuracy of the experiment?

ANSWERS ⟫

1 a $2H^+ + 2e^- \longrightarrow H_2$

   b The equation tells us that two moles of electrons are needed to make one mole of hydrogen. That is, two faradays, or $2 \times 96\,500 = 193\,000\,C$.

2 The measuring cylinder should be placed over the right-hand electrode as this is the negative one. The H^+ ions need to gain electrons to become hydrogen gas. These are provided by the negative electrode.

3 If the molar volume (volume occupied by one mole) of hydrogen is 24 litres mol^{-1}, then $24\,cm^3$ is

   $$\frac{24}{24\,000} = 0.001 \text{ mole.}$$

   This means that 0.002 mole of electrons (faradays) are required to form this volume of gas.

   The number of coulombs required $= 0.002 \times 96\,500 = 193$.

4 The variable resistor is used to ensure that the current flow is constant.

***Exam* tip:** These questions are based on the PPA 'Quantitative Electrolysis'. The main sources of error in this experiment are errors in reading the volume of gas in the measuring cylinder and fluctuations in current. The use of a variable resistor eliminates most of the current fluctuation. It is also suggested that current is allowed to flow for a minute or so before the measuring cylinder is put in place, to allow the electrode (carbon) to be saturated with hydrogen. Timing starts once the cylinder is in place.

Types of radiation (1)

Q1
a   What is the neutron to proton ratio in the isotope $^{210}_{84}Po$?
b   If the neutron to proton ratio is too large what type of radiation results?
c   If the neutron to proton ratio is too small what type of radiation results?

Q2   Answer these questions about alpha radiation.

a   What does an alpha particle consist of?
b   What is the range, in air, of alpha particles?
c   What kind of material will block alpha particles?

Q3   Answer the same questions as in Q2, but with reference to beta radiation.

Q4   Answer the same questions as in Q2, but with reference to gamma radiation.

ANSWERS ▶▶

1 a The ratio is 126:84.

b If the neutron to proton ratio is too large, a neutron will disintegrate to form a proton and an electron. The electron is emitted as a beta particle.

c If the neutron to proton ratio is too small, an alpha particle is emitted.

2 a An alpha particle comprises two protons and two neutrons.

b In air, alpha particles can travel a few centimetres.

c They can be stopped by paper (or skin).

3 a A beta particle is an electron.

b In air, a beta particle can travel a few metres.

c They can be stopped by a thin sheet of aluminium.

4 a Gamma rays are electromagnetic radiation (like X-rays, but more energetic).

b In air, gamma rays can travel many kilometres.

c They can be stopped by thick lead or concrete.

***Exam* tip:** Since the heavier atoms tend to contain more neutrons than protons (as in polonium above, with 126 neutrons to 84 protons) the loss of an alpha particle represents the loss of a greater share of protons than of neutrons. This increases the neutron to proton ratio. An alpha particle is identical to a helium nucleus.

Types of radiation (2)

Q1 **a** When beta radiation occurs, what happens to the atomic number?
b When beta radiation occurs, what happens to the mass number?

Q2 **a** When alpha radiation occurs, what happens to the atomic number?
b When alpha radiation occurs, what happens to the mass number?

Q3 Balance the following nuclear equation, giving values for x and y:

$$^{242}_{94}Pu \longrightarrow {}^{y}_{x}U + {}^{4}_{2}He$$

Q4 Balance the following nuclear equation, giving values for x and y:

$$^{45}_{20}Ca \longrightarrow {}^{y}_{x}Sc + {}^{0}_{-1}\beta$$

ANSWERS ▶▶

1 **a** The atomic number rises by one.

   b The mass number is unchanged.

2 **a** The atomic number decreases by two.

   b The mass number decreases by four.

3 $x = 92$; $y = 238$

4 $x = 21$; $y = 45$

Exam tip: In balancing nuclear equations it is necessary to make sure that the charge is balanced (atomic numbers) and that the mass is balanced (mass numbers). If the symbol for the product nuclide is not identified in the equation, it is necessary to remember that if the atomic number has changed, then the element has also changed so the appropriate symbol must be inserted.

Half-lives (1)

Q1 What is meant by the term 'half-life'?

Q2 What fraction remains of the original isotope after the following periods?

a one half-life
b two half-lives
c three half-lives

Q3 After 180 seconds the count rate from a radioisotope dropped from 512 count $min^{-1}g^{-1}$ to 8 counts $min^{-1}g^{-1}$. What is the half-life of the radioisotope?

Q4 Consider 0.1 g samples of pure Ra-226 and radium-226 oxide, RaO.

a How would the half-lives of the samples compare?
b How would the intensities of radiation of the samples compare?

ANSWERS ▶▶

1   The half-life is the time taken for the level of radioactivity to fall to half of its starting level.

2   **a** $\frac{1}{2}$

    b $\frac{1}{4}$

    c $\frac{1}{8}$

3   To get from 512 counts per minute to 8 counts per minute the radioactivity falls as follows

$$512 \longrightarrow 256 \longrightarrow 128 \longrightarrow 64 \longrightarrow 32 \longrightarrow 16 \longrightarrow 8$$

This means that six half-lives have passed. This has taken 180 seconds.

The half-life must be $\frac{180}{6} = 30$ seconds.

4   **a**   The half-life of a given isotope is constant, so whether the radium is present as pure radium or radium oxide makes no difference. The half-lives are the same.

    b   In 0.1 g of pure radium, all the atoms are radium. In 0.1 g of radium oxide, only half of the ions are radium; the other half are oxide ions. So the intensity of radiation from the radium oxide will be half that from the pure radium.

***Exam* tip:** The half-life of a particular isotope never changes. It is absolutely constant. Different isotopes have different half-lives. There is nothing that can be done to the isotope that can change its half-life in any way. It does not matter if the element is in the form of a solid, liquid or gas, or if it is converted into an oxide or a chloride. The half-life is always the same. The half-life is totally independent of its physical or chemical state.

Half-lives (2)

Q1 Sodium-24 is a beta emitter which decays to magnesium-24. A small cube of sodium-24 weighs 1 g.

   a What mass of sodium-24 remains after two half-lives have passed?
   b What will the mass of the cube be after two half-lives?

Q2 The table shows the level of radioactivity in a sample as time passes.

time (s)	0	5	10	15	25	30	42	60
counts (s⁻¹)	90	75	61	50	36	29	21	10

Use the data to estimate an approximate half-life for the isotope.

Q3 A radioisotope of bismuth was used to prepare a sample of bismuth nitrate and a sample of bismuth chloride.

How will the half-lives of these two compounds compare?

Q4 Explain why the following data cannot represent radioactive decay.

time (s)	count rate (s⁻¹)
0	64
40	32
60	16
70	8
75	4

ANSWERS ▶▶

1 a After two half-lives, 0.25 g of sodium will remain.

   b The mass of the cube remains at 1.0 g. Each atom of sodium that decays becomes an atom of magnesium with the same mass number.

2 It is not necessary to draw a graph to solve this. Inspection of the data shows that it takes about 18 to 20 minutes for the count rate to fall by a half. For example, between 10 and 30 seconds the rate falls from 61 to 29.

3 As already stated, the half-life of an isotope is a constant for that isotope. It cannot be changed by any chemical or physical process. Whether the bismuth is present as bismuth nitrate or bismuth chloride makes no difference. The half-life is the same.

4 The half-life of a given isotope is constant. It does not change during the course of radioactive decay even though the count rate changes. According to the data in the table the half-life starts out at 40 s, then falls successively to 20 seconds, 10 seconds and 5 seconds. As the half-life is not constant this cannot represent radioactive decay.

***Exam* tip:** In beta decay, the sum of the masses of the parent isotope and the daughter isotope remains constant because both isotopes have the same mass numbers.

As stated before, the half-life of an isotope is constant. If the half-life appears to change, then the process taking place cannot be radioactive decay.

Radioisotopes (1)

Q1 Radioactive carbon-14 is present in all living organisms. It is a beta emitter with a half-life of 5600 years. Why do all living organisms contain a steady level of the isotope?

Q2 Some wood from an ancient canoe was found to have a level of radioactivity at a quarter of the intensity of radiation in living wood. How old was the canoe?

Q3 Explain why, although diamonds are pure carbon, they contain no C-14 and therefore cannot be dated.

Q4 Suggest why the method is not useful for dating objects that are only a few years old.

ANSWERS ⟩⟩

1 All plants and animals take in carbon from their environments. Animals do this in the form of food, plants as carbon dioxide. Since carbon dioxide contains some C-14, all plants contain it, and so do all animals, as their food comes from plants or animals that eat plants.

2 As soon as a plant or animal dies, it stops taking in carbon, and the C-14 level in the organism begins to fall as the C-14 decays and is not replaced. When the level of radiation falls to a quarter of its original level, two half-lives have passed, a period of 11 200 years.

3 Diamonds are millions of years old. So many half-lives have passed that they now contain no C-14.

4 If the object is not very old, the difference between the level of radioactivity in a living organism and in the object will be too small to give an accurate estimate of its age.

***Exam* tip:** Neutrons entering the upper atmosphere react with atoms of nitrogen-14 to form carbon-14. This carbon combines with oxygen to give carbon dioxide, which is incorporated into plants. All food chains start with plants, so the C-14 enters all living things. It is steadily lost by excretion and other processes so that a steady level exists in all living things. On death, it is no longer taken in, so the level of radioactivity begins to fall.

70 Radioisotopes (2)

1 The isotope has a long half-life so that the smoke detector can operate for prolonged periods of time. Alpha radiation can travel only a few centimetres in air, so there is no risk of being affected by radiation. The alpha particles cannot escape from the detector.

2 **a** Steel 10 cm thick will not allow the passage of alpha or beta particles. Gamma radiation is much more penetrating.

   b The half-life must be reasonably long so that the device can operate for long periods without changing the radioactive source. In addition, if the half-life is too short, the intensity of radiation will fall and give incorrect information about the thickness.

3 **a** If used within the body, the half-life must be as short as is compatible with the isotope carrying out its task so as to avoid unnecessary damage.

   b Alpha radiation is highly damaging to cells and has a very short range. In any case, iodine has no alpha-emitting isotopes.

   c It is unlikely, as xenon is highly unreactive.

4 **a** Nuclear power does not produce carbon dioxide.

   b There is a risk of escape of radioactive material, and disposal of waste is difficult.

Exam tip: Applications of radioisotopes must take account of half-life, penetrating power of the radiation and risks associated with the isotope. Usually, for medical purposes short half-lives are favoured; longer half-lives are favoured for industrial applications.

Radioisotopes (2)

Q1 Many smoke detectors incorporate a small quantity of americium-241 (half-life 432 years). This causes the air in the detector to ionise and it can then conduct electricity. Smoke obstructs the alpha particles and the conductivity changes, activating the alarm.

Give two reasons why this isotope is suited to such a purpose.

Q2 In industry, cobalt-60 can be used to monitor the thickness of a product, which is rolled out in thin strips, such as steel. The amount of radioactivity that passes through the strip depends on how thick the strip is.

   a What type of emitter (alpha, beta or gamma) would be suitable for monitoring the thickness of a steel strip about 10 cm thick?
   b Why is it important that the half-life of the isotope is not too short?

Q3 Iodine-131 (half-life eight days) is used to treat enlargement of the thyroid gland. Iodine concentrates in the gland (which uses iodine to make an important hormone) and the beta particles destroy some of the cells.

   a Suggest what makes the half-life suitable for this application.
   b Suggest why an alpha emitter is not used.
   c Iodine-131 decays to xenon-131. Could this cause harm?

Q4 **a** Give an example of a major advantage of nuclear power.
   b Give an example of a major disadvantage of nuclear power.

ANSWERS ▶▶